JEAN McNEIL

Ice Diaries
AN ANTARCTIC MEMOIR

ALSO BY JEAN McNEIL

From the Library of Graham Greene
Hunting Down Home
Nights in a Foreign Country
Private View
The Interpreter of Silences
The Ice Lovers
Night Orders: Poems from Antarctica and the Arctic

CONTENTS

Note: This is a memoir and in order to protect others' privacy many people in this book are creations — characters, in other words. I have used first and last names only when they represent an actual person. Base R and some other Antarctic place names used here are pseudonyms; however, the names of landmasses and bodies of water are those you will find on a map.

"Snow mountains, more than sea or sky, serve as a mirror to one's own true being, utterly still, a void, an Emptiness without life or sound that carries in Itself all life, all sound. Yet as long as I remain an 'I' who is conscious of the void and stands apart from it, there will remain a snow mist on the mirror."

Peter Matthiessen, The Snow Leopard

"I think sleeping is a waste of time."

Max, aboard the James Clark Ross, Drake Passage, Southern Ocean

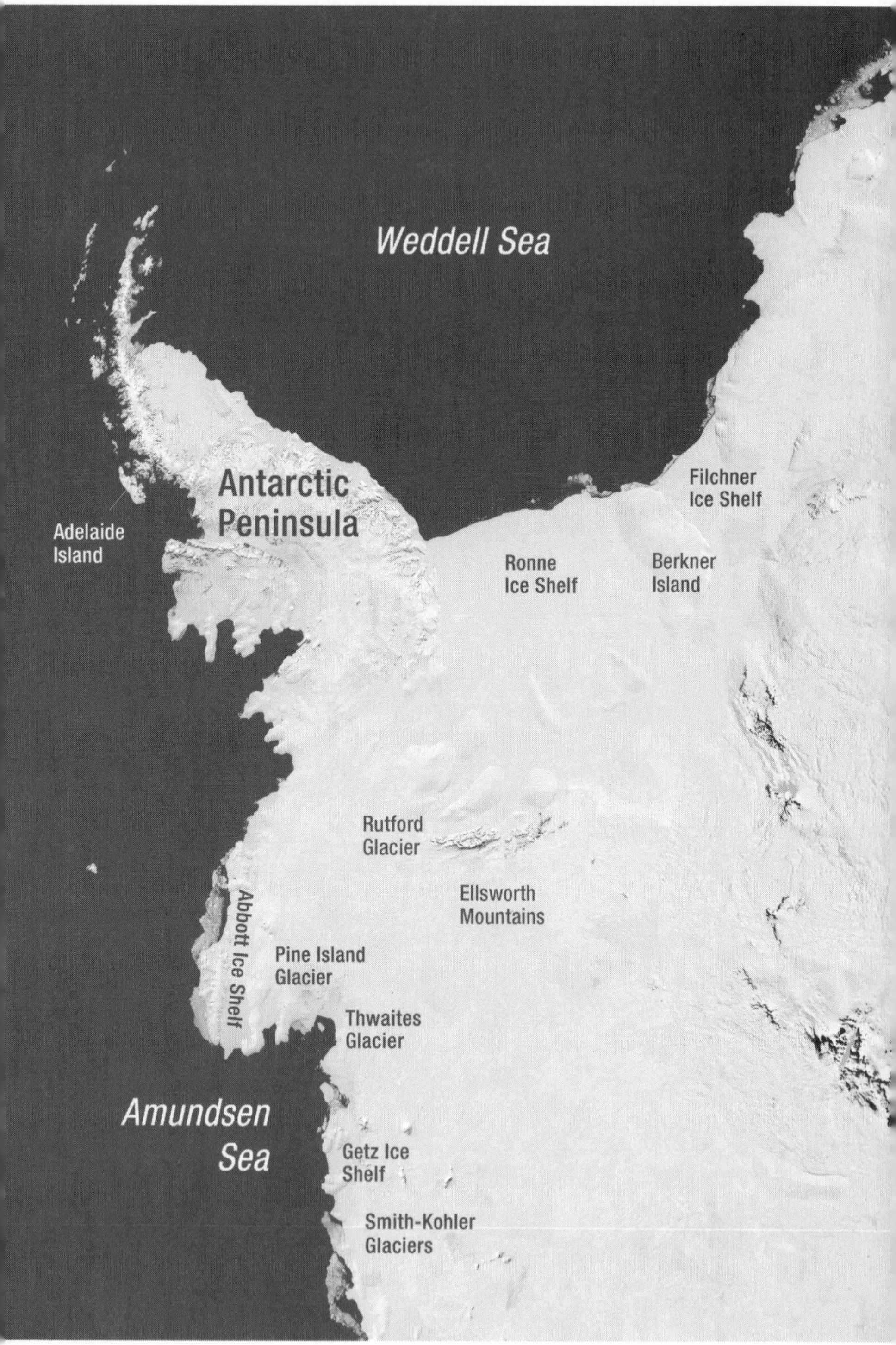

Weddell Sea

Antarctic
Peninsula

Adelaide
Island

Filchner
Ice Shelf

Ronne
Ice Shelf

Berkner
Island

Rutford
Glacier

Ellsworth
Mountains

Abbott Ice Shelf

Pine Island
Glacier

Thwaites
Glacier

Amundsen
Sea

Getz Ice
Shelf

Smith-Kohler
Glaciers

INTRODUCTION

It was Simon, the base commander that summer at Base R, who first told me about it. A strange little book was tucked away in the base library, he said, a list of all the different forms of ice, written for mariners, in the tradition of the identification manuals carried on ships: those glossaries of knots, or wind, or seafaring terms.

I'd encountered ice terms on the ship, a lollipop red-and-white polar research vessel which had conveyed us to the end of the world: I knew about frost flowers, polynyas, growlers — all words for the kinds of ice you encounter at sea, and had taken an instant shine to these gnarled, revelatory terms. How many more kinds of ice could there be?

After some rummaging in the dustless library (no mites means no dust — the Antarctic is heaven for neat freaks), I found the book. It was a thin paperback with stiff blue paper covers, worn but not "foxed," as they say in the book trade — those singed edges and brown liver spots which accumulate around the edges of papers and bindings — because there was no moisture in the air to fox them.

A Glossary of Ice Terms. The text was typed out in the blocky Courier of old typewriters. *Ablation, black ice, candle ice, first-year ice, frazil ice, glass ice, growler, hummock, ice gruel, pressure ice, rotten ice, sastrugi, serac, stamukha, tarn, winter ice*. The glossary listed over sixty ice words, in English. (There was a separate annex for Greenlandic terms.) There would be more to learn about ice than I had thought.

I signed the book out of the library using the outmoded index card system — a certain John Struthers's name, with the date 1978 and a book named *God's Mammoth Tasks* still shone in bright blue ink on the register, as if it had been signed only that morning. The slim ice glossary sat on my desk in the two offices I was assigned, Lab 7 and Lab 5, during the months I was in the Antarctic; by the time I left I had memorized the contents as you would conjugate verbs, committing to memory a language I never expected would be mine.

Ice has a life cycle, just as we do. We talk of it being born when an iceberg calves from a glacier, of its living and its dying when it melts. But actually ice is immortal; it never quite dies but is reincarnated, through melt, into water, into vapour.

The Antarctic is by far the largest accumulation of ice on the earth. The Ice, as the continent is sometimes referred to — a term of affection — is, along with Greenland, the most complete frozen archive of our planet's past. It is also an oracle, however reluctant or accidental. Through the chemical residues it traps, ice provides a precise record of the atmospheric past, and in particular how the planet has responded to past episodes of warming and cooling. Through analyzing this data, scientists become augurers: they can offer a likely scenario of how climate cycles and gaseous emissions will affect the future temperature of the planet.

Ice has also long been associated with another kind of divination: crystallomancy. The crystal itself has been used for future-telling through what in the eighteenth and nineteenth centuries was called scrying — looking into an object to divine a likely outcome. If we peered into it, mystics of the era believed, ice would reveal our future. Whether molecular or esoteric, ice promises revelation.

In the last ice age, the great ice sheets of the northern hemisphere were called the Laurentide, Fennoscandian, and Greenlandic and, in the southern hemisphere, the West and East Antarctic sheets. Of these, only Greenland and the Antarctic still exist. We remain in an ice age, but in a phase called an interglacial, when the ice sheets do not advance. This is an important point: shrinkage is not the key to an interglacial, rather lack of expansion is. So it has been for much of the planet's history. The next ice age was expected to begin sometime in the next ten thousand years, due to naturally occurring cyclotherms — subtle alterations in the Earth's temperature.

But now for the first time in the planet's long history, this natural cycle has been disrupted by man, and it is possible the polar ice sheets which wait so patiently for expansion may never again begin their advance. The Antarctic Peninsula, where I was based for several months, is an apt place to contemplate this interruption; the peninsula is the fastest-warming landmass in the world; at 2.5 degrees Celsius each year, it is heating up at approximately six times the global average rate.

The realm of ice is a mysterious world, underpinned by several persistent and little-publicized scientific mysteries, even conundrums. Before I went to Antarctica I knew little of the cryosphere, that sphere not of tears but of ice, apart from having grown up in a cold, still-glacial landscape. People from my background — North American welfare class — do not generally go to the Antarctic, or at least not the British Antarctic, and certainly not as writers. I had never dreamed of going there. The majority of people who go to Antarctica are scientists who work in the cryospheric disciplines: glaciology, oceanography, marine biology, terrestrial biology, atmospheric chemistry, physics, and astronomy, along with the engineers, technicians, mountaineers, pilots, and mariners who support the science programme on the frozen continent and who are the modern-day equivalent of polar explorers, carrying their cold studs of knowledge of the fickleness of ice, the harshness of the light, the topsy-turvy world which awaits there, where all your expectations and beliefs of how the planet ought to behave are overturned as the sun performs strange revolutions in the sky.

In the years I have spent teetering between the tips of the world since my first trip to the Antarctic, years spent in spells in Greenland, Svalbard, Iceland, Norway, the Falkland Islands, or shuttling between them on ocean-going research ships, everyone — scientists, seamen, pilots, electricians alike — has spoken of Antarctica's unique cultish allure. "Once you've lived there for any period of time, you can't get the place out of your head," Tim, an eloquent English builder

I met on base, told me. The Antarctic has been described by the Australian writer Thomas Keneally as not so much a physical place as "another state of being." The white tractor beam it exerts upon the consciousness of people who have been there, pulling them back, operates on a power beyond its obvious allures — its cold charisma, its pristine wildness. Could the lump of rock and ice that ballasts our planet at the bottom have qualities nowhere else on earth possesses? Could it be something akin to a consciousness? There, I felt the press of an external consciousness on mine, and not for the first time; as it turns out I had felt it as a child growing up in a cold corner of Canada's eastern seaboard, but had forgotten it.

Ice Diaries completes my trio of books on the polar regions. It is not a systematic account or explanation of the phenomenon of climate change or of the science behind it — other books, some of which I quote here, have done that far better. It is a less ambitious artifact: a witness statement, a travel narrative and a diary of a journey to the most inaccessible of the seven continents. It is also a study in the thermodynamics of our lives. Like the earth we also pass through cycles of cold, hot, and the middling temperate zones in between. Some years are spent in a cryogenic slumber. Some are incendiary and dissolve everything we thought was real — our homes, partners, families. How we weather these personal climate changes may determine the course of our lives.

Memoirs thrive on an overheated currency of self-transformation along the lines of *I was profoundly changed by this experience*. I'm not sure I can deliver on that promise with this book. Although I do go excavating for signs of change in my life, even as I live it; writers are avid archaeologists of the present. I find that after years in the polar regions my inner landscape is an altered state. *Things flourish in cold in a way they would never do in heat; you have made a career out of your sadness; I don't say goodbye, I don't look back; we aren't, after all, very good at imagining the future*: these are more than phrases uttered by those I met in Antarctica — they have a capacious intent to their tight explanations. They mean more than they appear to. They haunt me, still, which is a way of saying they have stayed with me, without resolution.

What is it about journeys? Aristotle said it best: "In the beginning, everything is possible. In the middle, one or two outcomes are likely. The end is inevitable." Aristotle was writing about narrative, rather than travel, but the two have much in common. A journey has a clean narrative arc that most of our lives — those 25,000 or so days between the gaunt endpoints of birth and death — lack. There is a beginning, and an end, and in between is what happens.

But journeys are not so simple, in my experience. My seven-year-long journey through ice was dogged by the future. During these years, in a way that would come to a dramatic culmination, I became more alert to what was going to happen. I took up scrying of a sort, and entered into the perilous world of trying to read the intentions of the future. I learned human beings cannot take the future, just as they cannot absorb much truth. The future and the truth are inexplicably intertwined, I discovered, so that they might be the same substance. At the same time, the polar ice caps have become synonymous with a particular vision of the future: of human civilization ruined, or at the very least profoundly altered, by climate change.

I thought I knew extreme cold from a hardscrabble upbringing on Canada's eastern seaboard. I thought about those years as little as possible; it was as if they had never happened. In going to the Antarctic, I believed I was journeying to a completely different place. The only thing it would have in common with my early life was that it would be cold. I might remember how to put on snow boots after fifteen winters in balmy England; I might have the advantage of knowing the signs of frostbite and how to prevent it. But that would be it. I was going to the Antarctic less as an individual with a past than as a writer, an official envoy from the future, to see what I would make of it.

The British Antarctic Survey (BAS)/Arts Council England International Artists and Writers residencies existed from 1998 until 2010; I was lucky enough to be part of this programme, which sent many talented visual artists, and a few writers, to the continent. We didn't have to sign a contract or otherwise pledge what we would do. We were understood to be witnesses. The expectation was that our resulting work would further the public communication of science, and in particular, of climate change science. We would make abstract entities and numbers appreciable to a more general audience than scientists could reach, through humanizing and personalizing them. People would care more: about science, about the polar regions, about climate change. As the English nature writer Robert Macfarlane has written, "We will not save what we do not love."

This undoubtedly happened, through the work of the writers and artists who were deployed to the continent. But writers are unreliable communicators. They bristle when told what they must say, what they should see. Not that BAS did, to its credit. The fault is with the sensibility of the writer. We automatically set about sabotaging the official line, consciously or unconsciously, even when it barely exists. We can't resist it, because one of the tasks of being a writer is to examine reality as it is shipped to us, preordered and prefabricated, for our consumption: what I call "consensus reality" — what everyone agrees, for convenience's sake, will be real. But look underneath the surface of the moment, and you find a roiling tension, not unlike the southern Atlantic Ocean in autumn and winter: a dynamic conflict so complete and vast it is hard to believe it remains intact. We are constantly on the verge of being torn apart by our realizations, and that is one reason why we deny them.

This book dramatizes what many readers may think of as science, but actually the science — information, observation — is indivisible in my mind from what actually happened, and what I felt. They are a single entity, like water, or even ice. To not have the science would condemn this book to solipsism. In Antarctica, information, experience, and endeavour are welded together; the Antarctic is a giant outdoor laboratory. Apart from a few high-priced tourist adventures, the continent is completely dedicated to science.

Science is a way of seeing things clearly, a process of revelation. A writer tries to see things clearly too, to perceive immutable truths that lie beneath the surface of the decoy we take to be reality. This can be an act of revelation but also of sabotage, because it might upset other vested interests in curating the truth. To discover these truths, things normally concealed — to the self as much as to others — must be brought to light. In the Antarctic I first learned that the original meaning of apocalypse (*apokalupsis* in Greek) is to uncover, to reveal.

I did learn a great deal about the science of climate change in Antarctica. However, climate is a flexible concept, and the book is an exploration of an inner as much as an outer landscape, and an account of the emotional effect of my rediscovery of winter that year — in essence, a kind of self-ambush. There is a winter of the heart, and winter of the mind, and these can be more chilly than any external temperature you might encounter. This book probes winter as a concept and an experience. Now that I live for part of each year in a tropical monoseason on Kenya's Indian Ocean, it seems to me a necessary season, a fallow lying-in, of regeneration and renewal of trust. Winter people are resolute and resourceful. We need these qualities to get us through lean times.

I do miss winter, the season of snow and ice, and the inner fortitude it summons. We could all learn from its austere glamour. What follows is an account of a life spent in ice of one description or another, and what my fascination with cold has bequeathed to me, and to the people who made me, as well as an account of a voyage of discovery — not of new lands or riches, but an interior lucre I was about to discover for myself: how the mind, the imagination, and the heart can be set on fire by ice.

PART ONE
RUNNING OUT *of* NIGHT

1. THE CAGE

sea smoke

Steam and fog over the sea formed by very cold air moving over warmer water, typically in the Antarctic in spring, when sea ice is thinning or disappearing, but land temperatures remain low.

MARCH 25TH

Before the cold, heat. I am back in the Cage. I never expected to see this place again after my pit stop here on my way back from the Antarctic two years ago. But I'm back, flying not to the white continent but to the Falkland Islands with the Royal Air Force. So close, but yet so far.

We sit at picnic tables sipping weak, salty coffee made with the island's desalinated water. I eavesdrop on tight-lipped conversations spiked with words like *deployment* and *operations*. While the officers sit in their neat pressed chinos and designer shirts, the squaddies stand around in boisterous groups, arms crossed over their chests. I see paunches, tattoos, outdoor-rugged footwear, and a high bald-head count. We are all listless. We have made that hasty transition, enabled by the age of air travel, from cold to heat in eight hours. Our eyes squint at the affront of light. Our backs sweat. We read magazines, or stare at some gadget or other, try to get signals on our mobile phones, even though everyone knows there is no signal on Ascension.

This is my second time on the RAF express. I find I've joined the Army again, although I can't remember enlisting. Ascension is a military airstrip, so any rules governing civil aviation and passenger rights can be smoothly cancelled. While the aircraft is refuelled we are all locked under guard in "the Cage" — this is how everyone refers to this half–indoor departure lounge, half–outdoor picnic area shaded by a corrugated tin roof and sealed with wire fencing.

Warning signs are wired to the Cage: *THE DANGERS OF FODs* one sign blares, followed by childlike drawings of penknives, corkscrews — Foreign Object Debris. Another sign says *Photography Prohibited*, so I immediately take a few of the unmarked aircraft on the tarmac (there is no proof that renditions flights have landed on Ascension, but it isn't completely beyond the pale, as only the US and British military really know what happens here): murkily incriminating images that look like they belong on the Amnesty International website.

It's back, the childish rebellion. Every time I am in an authority structure, I want to take it down, or at least disobey. But also it's reassuring to be back in a world whose workings I understand.

The scrawny palm trees rasp in the wind, just beyond the fence. A bougainvillea bush leans toward us. If I could only cut one of the fence chain-links, I could pick a flower.

I'd last been on Ascension two years before. Like many places I've been, I never expected to return. The island is not unlike the Antarctic: unless you are RAF, a Foreign and Commonwealth Office appointee, or an Antarctic scientist, everyone says you only get one shot at it.

Ascension is part of the quirky commute from the UK to the Falkland Islands and British Antarctic Territory. Before it was taken over by the UK and US military to be used as a mid-oceanic satellite/runway, its main use was as the Atlantic Relay Station for the BBC World Service, broadcasting to West Africa via vast electromagnetic installations that still stretch across a cindered lava field on the southern tip of the island. Nowadays it functions as a refuelling and spy station for the British and American governments. The island bristles with so many antennae, satellite domes, and giant wire contraptions like outsized dream catchers, I really did feel my cells buzz with microwaves.

Along with Tristan da Cunha and Easter Island, Ascension is also one of the most remote islands in the world. It lies nearly eight degrees south of the equator, on longitude 14°36' W, a desolate meridian only shared, roughly, with Dakar, Senegal, and Tristan da Cunha. It is almost midway between Africa (1,600 kilometres away) and South America (1,400 kilometres away). The nearest landfall from Ascension is St. Helena, 1,300 kilometres to the southwest, and we had another 6,000 kilometres to fly to the Falklands. As I sat in the Cage that March morning, I wasn't sure those distances defined anything at all. We were just very, very far away from anywhere else.

Two and a half years before, at Conference, as the yearly predeployment British Antarctic Survey gathering is called, I'd met a meteorologist who had done a six-month tour of duty on Ascension some years back. We stood clutching the stems of wineglasses in the cavernous dining hall at Girton College. I took to this met man: he was mischievous and slightly dishevelled. He wore professorial glasses and a striped polo shirt. He swayed in a way that suggested the glass in his hand was not his first of the evening.

"What goes on there?" I asked.

"You know, even though I was forecasting the weather for the UK and US military, I didn't have security clearance." He paused. "But there were flights nobody knows about coming in from the States to Botswana, Kenya, and Senegal."

"*Botswana?*"

"Apparently. I didn't even know Botswana had an army."

He proceeded to tell me another few choice anecdotes about the military ops on the island. "And now," he said, doing a little sashay of his hips and taking a step back from me, "I'm going to have to shoot you."

My last trip to Ascension was two years before I landed up back in the Cage that March morning. Then, I'd spent five days on the island as part of my homeward journey from Antarctica, out of curiosity more than anything. "You should definitely do a layover there," Paul, the director of the communications division at BAS, had advised. "It may be your only chance to see it. In fact you need permission from the governor to visit, but we can sort that out for you."

On Ascension I hired a car and drove around the island's forty kilometres of roadage in circles, in spirals. There were hardly any places on the island apart from military bases. The shabby conglomerations that did exist were called One Boat and Two Boat villages — Three Boats, it seemed, did not exist. I went to the gym at the RAF base, Travellers Hill, which looked like the summer camps I'd attended in Canada as a child, with flimsy wooden huts and beach towels hanging over the rails. Then I would drive down to the altogether more solid-looking American base, swinging by Wideawake Airfield on my way back into town, back to One Boat, then out to the BBC World Service radio transmitters with their signs warning of radiation. I convinced myself to drive up Green Mountain, the rainforested peak, and the only truly green landscape on the island's ninety-eight square kilometres. I wound around the serpentining road, passing through eerily quiet pine forests full of giant boulders, until a lava flow blocked my way. If Ascension ever got tired of being a spy station it could make a good living as a film set for movies with titles like *Volcano Apocalypse*. Certainly this was what the genesis of earth must have looked like: a wasteland of lava flow rubble and cinder cones from its forty-four volcanic craters.

I tried to parse the land; just as the island has no native population, it has no native species. The trees and flowers of Green Mountain were planted here beginning in 1850 when deliveries of nursery trees, flowers, and plants arrived from the hothouses of England, Argentina, and South Africa. Now the forest atop Green Mountain captures the moisture of passing clouds, enticing it to fall on its flanks as rain.

The greening of the top of Ascension's crater has been a great success; if it hadn't happened, the island would look like Lanzarote.

Buffeted by dry, sand-laden winds from the Sahara, it needed a long-term freightlift of trees to create enough rainfall for anything to grow. Ascension now has enough rainfall to support a menagerie of interestingly named species: the blushing snail, the bush cricket, babies' toes, and bastard gumwood.

I walked the sand-blown streets of the "capital" (actually a collection of jerry-built sheds) Georgetown, spooked by the island's remoteness, not caring whether I got sunburn on my Antarctic-white back. On the beach, giant green sea turtles tumbled ashore at night. I saw the tracks made by their flippers in the morning, the craters which they dug laboriously during the night to deposit their eggs. Nothing ever appeared on the horizon. The water and gas tankers permanently stationed offshore in case of shortages pirouetted on their anchors, and bored St. Helenan port guards slumped in pools of shade on the docks. Despite the heat, I couldn't even muster the energy to swim in the municipal pool by the docks — the only place to swim on the island, as the surf was too rough, patrolled by suctioning breakers, or fringed with knife-edged basalt reefs. If these were not deterrent enough, the island's waters fizzed with hammerhead sharks.

At night I sat on the floor of my room, accompanied by the bottle of Chilean shiraz I'd bought in the West Store in the Falklands, watching the single British Forces Broadcasting Services channel on an ancient TV. It wasn't only my shrieking anxiety that kept me awake, but the coughing St. Helenan man in the cubicle next to me, as well as the feral donkeys that roamed the streets of Georgetown, squealing their banshee hee-haw call all night.

The only other guests were St. Helenans waiting for the ship home. A group of ladies addressed me. Dressed in Sunday church attire — neat polka-dotted dresses and blue hats — they sat on the veranda, watching the wild donkeys wander down the road.

"Hello," they said amiably. "Are you married?"

"I — well." I decided to lie. "Yes."

"Oh," they grinned. "Good!"

I went to walk away, then returned. "Why did you ask me if I was married?"

"Well, that's what everyone wants to know on St. Helena."

I didn't say, You're not on St. Helena now. I only smiled and basked in their approval.

Now, sitting again in the Cage, drinking my salty coffee, I realized every man looked like Tom: checked shirt, outmoded jeans, shoes a cross between hiking boot and trainer, complicated watches on their wrists. Their balding, often greying hair sheared sharply at the neck, their faces speckled with age spots from flying too close to the sun.

Sometimes I actually see Tom, or think I do — in the street, in train stations, airports — when in reality I saw him last two years before, on the apron of the aircraft hangar at Base R. Another Antarctic ghost.

I struggle with an aspect of my life which I can only term phantasmal. People come and go inexplicably, like badly announced characters in Elizabethan drama. They stand rigid in the spotlights, compelling, seemingly eternal. Then they vanish. I don't kid myself I am innocent, nor do I think it's a particularly unusual experience of life — that people are transitory, unreliable, unfathomable.

Still, on that morning on Ascension, I felt the return of a sensation I had in the Antarctic during the months I spent there two years ago. It was the impress, faint but unmistakeable, of a larger, nameless intelligence. I was under its jurisdiction again. Outlandish places not subject to the usual safeties of life make us feel more aware of the vulnerability and arbitrariness of our existences. Here I was on an island with no mobile phone signal, no cash machines, no cash for that matter — apart from the raucous St. Helenan currency, with its blue wirebirds and pink anemones. I was in a place that supports the Global Positioning System that tells us where to find a petrol station on our mobile phones, but which the UK and US would keep secret from the rest of the world, if they could, in an age of satellites and flight tracker airport codes. I felt the newness of the land there; it was still working out its level of commitment to the planet. It could be sucked back down to the ocean floor in an implosive volcanic burp, and all us with it.

Start from the beginning. I've learned this as a novelist, and it's almost always good advice. But often a story begins in the middle, or even at the end. Just as the iceberg, once calved from the continent, revolves restlessly, caught in the gigantic gyre of the Antarctic Circumpolar Current, we also turn round and round. We are not caught in a single line, a narrative, a parable that lurches from A to B, nor a circle, but in a spiral.

An announcement came over the tannoy. A snappy voice issued our orders.

"Attention, everyone; would squadron leader Christian please report to the Air Movements NCO. Would passengers replane the aircraft in the following order: officers, civilians, followed by personnel."

On the RAF plane there was a new hierarchy, quite different from the über-businessmen or Home Counties Brahmins you see in first class on British Airways. Here the people in the front of the plane were the officers, while the squaddies and civilians travelled in steerage. There was a square for rank on the boarding pass (mine was "Ms"). Then there were the Falkland Islanders and the St. Helenans, who everyone calls "Saints." A couple of Saints stood in front of me, sleepy in the early morning, slumped against each other, their hands around each other's waists. This was their patch of the planet, our two-hour wait in the military lockup as unremarkable for them as a Heathrow transit lounge.

We filed toward the aircraft; I had the impression we all dragged our feet, reluctant to leave the tropical sun behind. In the southern hemisphere the planet had tipped into autumn. I could already feel its taciturn, enigmatic presence. Everyone in the British Antarctic crowd calls it simply "South." As if everything else plunked south of the equator were drowned in the dazzling reality of the ice continent that needs no name, only a cardinal direction.

I remembered what Tom told me about the continent's allure, the magnetic pull it exerts on you. We were sitting in the fuselage of the Twin Otter in a mild whiteout, waiting for the visibility to clear so that we could take off and return to base.

"I've been working South for twelve years now," he said. "In the beginning, when I first flew down here, I couldn't get it out of my head. In the winter I'd be working in the Falklands, or flying in the UK, and suddenly I'd have to stop what I was doing. I'd just see it: a solid plain of white. And need it somehow. I always felt reassured, knowing I was coming back here at the start of every new season."

"And now?" I asked.

"I don't know." He shook his head. He looked . . . I don't know what to call the expression. Defeated. Possibly guilty. "Now I see only the deprivations. Now, once I leave, it doesn't seem real. Why do you imagine that is?"

"Maybe because your family isn't there. No one you really love is there."

"All I know," he said, "is that once I leave it now, it stays behind me."

Seated on the plane, an ancient DC-10 — a gas-guzzling 1970s model flown only by cargo companies these days, chartered by the RAF — I turned my face toward the day's first rays, and felt the ghostly appeal of a journey I was not supposed to be making. I was on my way to the Falkland Islands, where I would undertake a two-month-long writing residency.

But I was supposed to be somewhere else. Most people I knew thought I was at sea in the Southern Ocean on a research vessel. Behind me lay an abandoned trip, a strange lie, and a catastrophe which I had, to an extent, foreseen. I was flying to the ends of the earth over empty ocean on a creaky plane, so I listened to myself for signs of the turmoil I felt only weeks before in Cape Town, and which vaulted me off my intended path.

I no longer felt I knew who I was. This had occurred to me before — the possibility that we do not own our own bodies, even our souls, rather they are on loan to us from the destiny-givers to see what we do with them. As for character and identity, those constellations of sparks and instincts and fears which drive us, these might be a necessary fiction in which we largely believe.

Then, a very few times in life, we do something — take a sudden and inexplicable decision, make an avoidable mistake, are overtaken by an emotional breakdown — that shows us that we don't know who we are, or what we want, or even what we are going to do from one minute to the next. The illusion of the coherent self is spoiled, and we become strangers to ourselves, commuting between the mystery of our desires and the limits of our abilities.

Suddenly my limitations were just what they were: limiting. I hardly needed the mesh and wire with which the Royal Air Force had seen fit to enclose me. I was my own cage.

Some places do stay behind you. But others refuse to assume their rightful position on the linear timeline. These form islands in the river of time and in memory, persistent and opaque. There live people and events that happen over and over again, spiralling out beyond that which can be described as already experienced and so known; something about them is being worked out on a timescale far grander than the moment, or our individual lives. They are the past, but the future also.

Max, the climatologist I met in the Antarctic, told me how computer code gave him entry into the hidden pockets of the continents' past, so that he could make accurate predictions about the rate and intensity of the warming planet: "If you want to know the future," he said, his voice ringing with the bell-like certainty of the very young, "you must look into the past."

We roared up the four-kilometre-long runway, the fifty-something Texan stewardesses of the aircraft leasing company vibrating in their seats. Then we were aloft, and the air around us tightened with sudden cold. There would be ten hours of ocean and sky before we saw another scrap of land.

The plane banked, then righted itself over the brittle edges of the island, its shark-stocked waters. We were back in the troposphere. The ancient plane roared. Its nose pointed away from Africa, South America, toward a nullity. We were heading south.

2. A PROPHECY

rime

A white or milky opaque granular deposit of ice which forms on exposed objects at temperatures below the freezing point.

The taxi serpentined around roundabouts, past hulking gabled buildings, grey-brown, sandstone, russet, which poked out of woodlands and puffy trees and clearings like eternally reposing cranes. Cyclists streamed past, dressed in tweed and demure skirts, as if time had stood still, or maybe they were extras in a film, a biopic of Wittgenstein or Turing perhaps, or any one of the great minds who had flowered there.

Cambridge, of course. There is nowhere like it, with its ashy buildings and Waitroses where Nobel Prize–winning physicists queue to buy sushi.

We had an Indian summer that year. It was a warm day in early September. The train from London had passed through hot horizontal towns and fields so flat and neat, it was as if they'd been drawn with crayon.

I shared a taxi from the station with Suzanne, a Cornish sculptor, lately a Londoner like me, who was to be the visual artist that year on the programme. She would make ceramic sculptures not of ice but of the outbuildings and electrical installations, the VSAT domes and the mechanized snow-movers found there — the unlovely detritus of heavy machinery. "Post-industrial aesthetic," she explained.

We didn't know what to expect. Only that we'd been told any trip with the British Antarctic Survey starts at a red brick Cambridge college. Here, as summer draws to a close, hundreds of people with disparate jobs and nationalities converge for a three-day-long conference. They have only one thing in common: Antarctica.

The British Antarctic Survey has one of the longest-running programmes of Antarctic science among the thirty-odd nations that maintain research programmes on the continent. In part this endeavour is a hangover from the great British era of polar exploration, led by the Edwardian hero-rivals of Antarctic discovery, Robert Falcon Scott and Ernest Shackleton. Nowadays BAS is purely scientific, although its prowess in earth sciences and physics has been supported by Britain's long expertise in mounting Antarctic expeditions.

Girton College had been hosting BAS' summer conference for as long as anyone could remember. As a building, it was strict and imposing. To escape its endless hallways and grumpy portraits of past dons, I went for a run around its perimeter that first day, and had a strange experience. As I was running, a mass of air, person-sized, came up behind me and whispered something across my neck. I started and looked around me, but there was no one there — no other runner, no would-be assailant. I remembered the gust of air that accompanied that voice long afterward. It had mass; it was electric but invisible. Girton must have its share of ghosts, I thought, particularly the ghosts of unhappy young women married off after obtaining a token education — for most of its history it was a women-only college.

In its basement bar those nights, young men and women wearing fleeces honed their pool-playing skills and drank pitcher-sized mugs of beer. It was bracing to suddenly be in the company of so many young people. The atmosphere was boisterous, heady. Everyone looked the picture of health; most of the field assistants, as the mountaineers who accompanied scientists on expeditions were called, had just descended a mountain to attend Conference. After it was over, they would return to Snowdonia or the Cairngorms to perch anew, waiting the waning of the days when they could trade northern hemisphere summits for their cousins in the Antarctic.

Suzanne and I hovered by the bar. We exchanged glances, thinking, Safety in numbers, even if the number is only two. We were already something of a double-act, reinforced by the fact that we were physically identical: small, dark-haired women. We were dressed normally but still came across as a couple of goths parachuted into a North Face catalogue.

But we were enlivened too. We would soon leave typical North London bohemian existences behind: raw battles over decent flats in Islington and Stoke Newington, negotiating gallery and literary agents' sales commissions, keeping a wary eye on whose book was excerpted on BBC or whose show was reviewed in *Art Monthly*. It was barefaced survival of another kind, where we came from. As we surveyed the crowd of hale, normal youngsters that night, we wondered what we'd got ourselves into.

"No one drifts into the Antarctic," Paul, the director of the information division and our contact for the Artists and Writers Programme, had told me when we'd first met three months previously. "You passed through the same selection process we apply to everyone — plumbers and pilots alike."

Up to four hundred people applied for every job going in the British Antarctic, Paul had told us. The pilots weeded themselves out; there were only a handful of people in the world who could, or would, fly in the conditions experienced there. The scientists — I was to be a scientist for logistical purposes in the Antarctic, and for a while basked in an unfamiliar sense of legitimacy — had to go through repeated rounds of funding applications. I was congratulated by those people I met for having passed a rigorous selection process in which my application had been examined by committees I had never met.

Paul himself was an example of the Renaissance men and women who I would come to meet in the Antarctic — a scientist equally well versed in the arts, refined, cosmopolitan, and utterly devoted to the continent-sized chunk of ice at the bottom of the world. He had the unusual intensity and self-possession I would eventually encounter in many Antarctic veterans. They struck me as people who had confronted the limits of themselves, in a way that one witnessed in people who had been through wars or other species of conflict.

For the moment, at Conference it was evident that everyone in the Antarctic world had fought to get there, and seemed to possess an almost messianic motivation. There was no cynicism and no entitlement, and this was refreshing.

At one of the Conference soirées, I ended up talking to Mathieu, a French glaciologist who had worked with BAS for over a decade. He had been drafted in by BAS to give a presentation at Conference on Arctic versus Antarctic ice dynamics. Mathieu had done twelve seasons south; in the summers he went to the Arctic to work.

"I am an Arctic tern," he announced. I scrutinized his face for signs of polar faceburn. The polar regions were hard on the skin, I'd been warned. The combination of the extreme dryness and the intense UVA and UVB rays from the polar sun could put five years on your face in a single three-month summer season.

Mathieu looked about twelve. He wore a blue shirt over a white undershirt and beige chinos — it seemed this was something of a uniform among the scientists.

"But I haven't actually been south in ten years. I did my doctorate and post-doctoral research there, but then I defected to the north," he said.

"How are they different, the Arctic and Antarctic?"

Mathieu's head rotated — not quite a shake, not quite a gesture of admonition. I had asked an impossible question.

Mathieu was measuring his words, possibly calculating them for the imbecile he had in front of him, who had been to neither pole. "They are both cold, obviously. But even there, no comparison is possible."

He peered at me again, with his level, brown-eyed gaze. I would get used to this feeling, in the months to come — of shame in my relative ignorance, at being the least-informed person in the room. I knew about their names, at least. The antithesis between the two regions is rooted in Greek: *Arktos*, from the Greek *arktikos*, meaning northern, has its origin in the Greek name for the constellation Ursa Major, the Great Bear; and *Antarktikos* is that which is the opposite to *Arktos*. This mirroring extends further than names. As the American environmental historian Stephen Pyne observes in his magisterial history of the continent, *The Ice*, "The Arctic is a true ocean surrounded by continents; the Antarctic, a continent surrounded by oceans." In other aspects, too, they are radically different — their human history, their climates, their ocean currents, their types of ices and glaciological history.

"The Antarctic is far, far colder," Mathieu said. "You feel it like a blow, on the ice cap." He thumped his chest with his hand. "You can freeze your lungs, just by breathing. In the Arctic you have to go to Greenland to find that kind of cold. In the winter," he added. A new expression settled in Mathieu's eyes. Something clouded, but rhapsodic. "The Arctic is so alive!" He shook his head, as if he still could not believe it. "Foxes, snowdrops, moss, reindeer, people."

"And the Antarctic?"

"Dead. But compelling."

"How can something dead be compelling?"

He gave me an unreadable look which might have said, *You'll see*.

"Do you miss it?"

"I do," he said. "I think what I miss most is not what I expected."

"What's that?"

"It really is the most lethal place you can go," Mathieu said, shaking his head. "It will kill you like that!" He clicked his fingers.

"And that's what you miss?"

"It's the sensory deprivation. The Antarctic is unique — even the Arctic isn't as extreme. There is more sound there somehow. Even the wind sounds different. In Antarctica the wind doesn't blow, it *scours*. You stand looking at the ice sheet and you can hear nothing, nothing but the beating of your own heart." Again, he pounded his chest. "Sound and smell are gone, wiped out. That's two of the senses dealt with. What a relief."

I wondered if death-wish characters were attracted by the continent, or if the clinical deadness of the place provoked the desire to muffle the self, to disappear. History proves that the Antarctic draws unusual — quite possibly unstable — characters, and its relation to death is well documented.

"The Antarctic is completely Other," Paul had told me, when I'd visited BAS headquarters in Cambridge in May that year, shortly after being awarded the fellowship. "There really is nowhere on earth like it." His statement had an obvious, almost liturgical, ring. But also, something of Paul's fervour began to sink in to me. I started to form mental images, purloined from film and photographs, but which were also a projection of my own specific fears. I saw fields of white rippled by the cobalt seams of unseen crevasses. I had read Jon Krakauer's *Into Thin Air*. It had given me a sense of what it was like to vanish into one of those frozen fissures, to be eaten alive by the cold earth.

There is a long tradition of death in polar literature, although I hoped it wasn't obligatory. Before writing my application to the Artists and Writers Programme I had read the classics of explorer literature: *South: The Endurance Expedition* by Ernest Shackleton, *The Worst Journey in the World* by Apsley Cherry-Garrard, and a volume of Captain Scott's letters. Taken together, these narratives of men (they were all men, then) who had pitted themselves against its annihilating *froideur* were shaming compendiums of selfless toil and hardship and an effective discouragement against ever setting foot in the place.

While death seemed to be the default story and the only

permitted metaphor for the continent, its antithesis was also stirred. As Francis Spufford points out in his cultural history of Antarctic exploration, *I May Be Some Time*, while purporting to be about the external environment and the challenges the continent sets to your existence, as well as its regal indifference to the fate of any human being, these books are also about the consciousness of the explorers themselves. Explorer literature was affirming. Stalked by death, the continent elicited extraordinary feats of courage and endurance, pushing those penitents bent on conquering it beyond the boundaries of human suffering and into a near-beatific realm.

Reading these books, the question that occurred to me was less how these men managed to survive their physical ordeals, but about their emotional welfare afterward, if they lived. After such lives of triumph, despair, brotherliness, and grit so powerfully evoked by the writer-explorers of the day, how did those men return to normal life, I wondered, to the hedgerows and weepy hydrangeas of Edwardian England, the endless tea receptions with dowagers that would follow their inevitable lecture tours on their return. Had adrenaline not colonized their senses, simultaneously sharpening and muting them, so that it became impossible to calibrate their emotions to the milky scale of ordinary life?

As I read, another theme appeared in the pages of these gruelling catalogues of mishap, bad weather, bad judgment, confluences of all these, and sorrow: luck. Death and luck — not a reassuring duo. I might have trusted death more than luck, at that impasse in my life. That these two unreliable spectres should duel it out in the blank heart of the continent's whiteness gave me pause.

I leaned against a red brick wall, listening to the genial ship's officers regale us with seasickness remedies (ginger biscuits). My journey to the Antarctic was then two months away, still notional, and cancellable. This place, uniquely in the world perhaps, had the capacity to separate you from yourself. You would be divided, flensed into components you might not know you possessed. It was a journey of a different disorder, one which promised no intact return.

SEPTEMBER 2ND

My thirty-seventh birthday. I have completely forgotten about it. Text messages from friends say, *Where are you?* I sit alone at night in my college room. A blank bulletin board eyes me. Blu-Tack stains pucker the walls. It will soon be occupied by an undergraduate, counting his or her fortune to be at the top university in the country.

I am trying to take stock of the situation. The boisterousness I observe in the college bar is in a class of its own, the kind of boosterism that generates pranks and cliques in equal measure. They are all pathologically exuberant, and I'm used to moroseness, a certain hedging of bets. My notes from today: *a unique society, likely unreplicated anywhere in the world, a marriage of sedentary thinkers glued to increasingly complex computer-modelling programmes and go-for-it adventurers.*

I am keeping a list of the occupations of the people I meet: carpenter, first officer (ship), chief pilot, glaciologist, atmospheric chemist, terrestrial biologist, chef, mechanic, JCB driver. Everyone talks to each other; there is no obvious class division between the logistics people and the scientists. This romance between science and logistics has a military feel to it, in part because of the contribution of the Royal Navy, in the form of their vessel *Endurance*, and the two certifiably reckless trainee Navy pilots who are here and who will spend the summer on base, as flight and communications assistants.

Why is everyone so thrilled, so motivated? I'm a writer; no one in my profession admits to anything as simplistic as certainty. Here, Antarctic veterans regale me in the bar. They are not telling stories, or not only. Their stories have an aspect of spiritual warning.

The Antarctic is an addiction, they say. No matter how sick of the privations of life there you thought you had become, it always tugs you back. At the end of the season, all you could think of was getting out. But then as soon as you do escape you find you have amnesia and you can only remember the good parts. Now all you want to do is get back.

The world will never look the same again, they say. Nothing is ever quite as real again, because you are not there. Your dreams become populated by vast lozenges of ice that turn into icebreakers, or that house red planes, like frozen aircraft hangars. But you will not be allowed back because you've had your shot at it, and there are others queuing behind you, and access to the continent is restricted. Even before you went there they said, *This is a once in a lifetime chance*, the word *once* taking on an alarming ring, like a medical condition. For years you will wake, sweaty and disoriented, from your dreams of ice and think, Why can't I go home?

Two days later the whirlwind that is the Antarctic Conference packed itself away. The Foreign and Commonwealth Office people were driven back to London in Jaguars, the officers rejoined their ship somewhere on the Humber River, the scientists went back to number-crunching in front of their computers.

Suzanne and I shared a taxi and a train back to King's Cross, to resume our London lives for two months until our marching orders came through and we joined one of the first BAS group flights in the third week of November.

"That was so much more fun than I thought it was going to be," Suzanne said. I agreed, although I was struck by a sense that artists and writers were, while valued, not a priority to the programme. My first inkling came while sitting next to the FCO polar regions representative at lunch on the first day.

"How closely are Britain's strategic interests in the Antarctic aligned with the policing of the sovereignty of the Falklands?" I asked. She looked at me as if the potato on the end of her fork had just grown a mouth and spoken. She did not deign to answer and spent the rest of the lunch talking to the deputy director of BAS, seated on her left.

"How about you?" Suzanne asked brightly.

"I don't know. It was —" I searched for a neutral word. "Overwhelming."

"What will your book be about?"

"I don't know. It depends on what happens there," I said.

She nodded, but I could tell she was not satisfied. I didn't feel I could reveal to her the full extent of blindness often required in writing fiction. It must seem haphazard and unwise to people who do not charge themselves with making up stories out of thin air. The one determinant factor of fiction is that there will be a narrative of cause and effect, enacted by characters, with a beginning, middle, and end. But I had no idea what this would be.

Outside the train window the same fields flashed by, only in the last three days they had somehow acquired a slight verticality and now tilted into the sky.

"What made you want to go to Antarctica?" Suzanne asked.

"I wanted a creative challenge, I suppose."

My answer was honest, but its honesty disguised another, more ambivalent stratum of truth. It is a writerly preoccupation, perhaps: to locate the moment when a spark is struck, when an idea occurs to you that will set in motion a train of events we will come to think of as our fate, our destiny, or simply how things turned out.

My enchantment with the polar regions began with a casual prophecy — one I would forget before remembering.

I see people standing around, dressed unusually. Snow, ice. Are they Eskimos? Are you going to write about the Arctic?

Denise was a friend of a friend, a very successful astrologer. She did the charts of the well-off of Holland Park, she advised CEOs of multinationals on the stock market. I never met her in person, but a friend ordered a chart from her to be drawn up for my birthday over a year before I went to the Antarctic. Denise rang me to say that the chart was finished, and that she would put it in the post. There were a few things she wanted to "run through" with me first, though.

"There's something I saw," Denise said. I could hear her ruminating on the other end of the line before she told me.

As she said it, I thought, I doubt it. I'd lived through twenty Canadian winters before coming to live in the UK. I'd had enough cold, thanks very much.

I said, "Are you sure?"

In her voice a tiny bell of annoyance sounded. Of course she was sure.

I was in a hotel in Rome when I received the email from the British Antarctic Survey. My trip to Italy had taken me on a slightly surreal and lonely circuit of universities and lectures — Siena, Trieste, and now Rome, where I talked to students about being a writer in cavernous university halls filled with a bronze spring sun. The Vatican had just elected a new pope — white smoke had been seen floating from the chimney in St. Peter's Square, only three kilometres from where I lectured.

Back at the hotel, I fell off my chair — or rather the chair declined to stay underneath me. In any case the wheels slipped on the polished marble and deposited me in a heap on the floor. Dapper Italians looked at me in blank shock.

Denise's words bloomed in my mind like a neglected flower. I'd forgotten what she'd said, but now I would never forget. That is one of the many problems with prophecy: once you know it, you can't unknow it. You either pitch yourself against it, or subject yourself to its diktat out of a vague sense of obligation.

I read the email from BAS again, as if the words might expire or begin to erase themselves at any moment. But they remained, etched into the screen. It was no longer an unhinged notion, a creative stab in the dark, an intuition. I was going to Antarctica.

3. THE IRON ISLANDS

brash ice

Floating ice rubble. Originates from sea-ice that is breaking up or commonly as debris from calving ice bergs or ice bergs that break up as part of their ongoing erosion. The wreckage of other forms of ice.

Late November. For the first time in my life I was handed a group airline ticket booked and paid for by someone else, my name only one of many on the manifest. Our route would take us to Madrid, then Santiago de Chile, where we would spend the night. The following day we were to take a hopscotch flight to Punta Arenas via Puerto Montt, finally pitching up somewhere called Mount Pleasant Airport.

At Heathrow we milled about in an undisciplined group, each of us lunging off to procure last minute toiletries or have their last cigarette for twenty hours. Suzanne was already in the Antarctic, having left two weeks before me, so I had met only one person in our group before: Paul, the head of the information division, who trotted after us, a bearded and distracted mother hen.

Gradually other likely suspects joined us. One man, his face mottled by burst capillaries, carried a bag with the organization's insignia. Then other burly men appeared, their heads shaven close. I sat next to a plumber who said two or three words to me, then folded his arm across his chest and stared at the back of the seat in front of him for the rest of the flight to Madrid.

In Madrid new faces joined us — Nils, an oceanography student from Norway; Emilia, also studying oceanography, from the Italian Antarctic Programme; and Max, who was from the Max Planck Institute for Meteorology in Hamburg, although he was Swiss — I couldn't remember whether he told me this, or Nils told me, or I inferred it. I didn't know it yet, but we would spend so much time together, all of us, that it would come to feel as if I had always known them, and would always, in a kind of glutinous and total knowledge of another person's existence I hadn't experienced since university.

Max sounded English — more English than the English, with crystalline, carefully acquired consonants. Emilia was small, dark-haired, pretty, and a little stiff. She flashed thin smiles. She had a sheltered look about her. She would shrink from the reckless camaraderie of strangers shoved together by circumstance. Nils was a more muted presence; he might be a couple of years older than Max. Nils publicly turned off his mobile phone, holding it up and waving it at us. "Bye bye, Mum and Dad, bye bye, Nadia — that's my girlfriend."

There would be no mobile coverage in the Falkland Islands, and it was likely our networks wouldn't be available in Chile, or would be too expensive to use. For some of those headed to the Antarctic on long-term contracts, that hour in Madrid airport was the last time a mobile phone would be of any use to them for two and a half years.

Max, Emilia, and Nils would be in the Antarctic world for six weeks and on the continent itself for only three or four days. They were all on secondments with their supervisors, whom they were joining on the ship. Nils was doing a project for his postdoctoral fellowship in Tromsø. Max had recently embarked on a PhD.

I realized that I was not often among people in their early twenties. We are herded into our age and peer groups by school, then university, then career, by friends and family, and what we might call "lifestyle choices." In Antarctica I would meet almost no one my age. I would discover that the Antarctic world was stocked with people in their twenties, and people in their forties, fifties, and beyond. The thirty-somethings were not banned, but they tended to have young families, and the Antarctic required long periods on the other side of the world with little communication.

I looked at them around the table with some envy: Max with his long legs splayed proprietorially, enigmatic Emilia, upright, friendly Nils. This voyage would provoke lifelong friendships between them, or perhaps more. Meanwhile my life had solidified into a pattern — work, interests, friends — that hadn't changed in some years. I wasn't sure if change was still possible.

Although short-term convulsions seemed to be a speciality of our destination. People went off the rails in the Antarctic — several people I talked to at Conference had said this, then illustrated it with lurid anecdotes. I had been warned by another writer I knew who had gone to the Antarctic with the American Antarctic Artists and Writers Programme. "They have a saying: *What happens on the ship, stays on the ship. What happens on base, stays on base.*" The Antarctic, the writer said, was a secret society. As soon as you were initiated into it, your social norms and duties were miraculously extinguished. People fell into catastrophic liaisons, got drunk and made a pass at their bosses, or students, or suddenly forgot they had a wife and three children at home. "The men, especially," she said, but that was hardly surprising — Antarctic society was still two-thirds male, on average. "It's quite entertaining."

NOVEMBER 26TH

Fourteen hours. It is good to be back in Latin America, even briefly. I fall asleep over the Amazon. I have a dream about a man. I don't know him. I am keeping him company. He is waiting for my friend Rebecca, who is beautiful. I remember only the marshy, compromised feeling of entertaining a man who is waiting for a more beautiful and seductive woman.

In the morning we fly over Salta. I think of Lucrecia Martel's film *La Ciénaga*, or *The Swamp*, which was filmed in the province, its lack of an establishing shot. You get lost within it, as confused as the characters who live in the house where it is set. This heightens the feeling of claustrophobia and significance. I want to do this in fiction, but how? To start from the inside, to begin inside the emotion. Instead of having to explain, to make a coherent fabric out of emotion and experience, which in reality refuse to knit themselves neatly together.

Max, Emilia, and Nils sit toward the back of the plane, plugged into films. When I go to the galley or the toilet, I pass them but they do not look up. We know we will be spending weeks in each other's company, perhaps more, so we measure our interactions carefully. I read the flight map for hours. La Serena, San Juan, Tucumán, El Salvador, San Luis, Córdoba, Jujuy, Juliaca, Viña del Mar, Copiapó. Names on the moving map rise up to meet us over the artificial horizon, then disappear in our wake.

NOVEMBER 27TH

Santiago. Nineteen-fifties window displays. Chaotic department stores. The Moneda Palace is actually very ugly. Military police drift around its perimeter, as if awaiting insurgents. But there are only tourists who transgress by stepping on the grass or into the fenced-off enclosures.

We have dinner together in a group. A scientist in our party — who? Didn't catch his name, I can't seem to retain anyone's names, even though we are only twenty people or so — telling me at dinner about "black smokers"; these are glass spheres raised from the ocean floor by underwater magma. He shows me a photograph. They are gothic vases, twisted yet somehow coherent. Sometimes they implode, he says, from the pressure. When they do, he says, they crumple like pieces of paper.

The following morning we began our long journey down the thin finger of the country, leaving the vine-tangled valleys of the Chilean winelands behind us, flying over garrisons of spruce, until we arrived on a wind-scoured runway in Punta Arenas.

There, in Punta Arenas' airport, where the sword-like light of the extreme southern hemisphere pierced the departure lounge, we began to feel the Antarctic's presence. There was an empty note to the sky. But it wasn't an emptiness of vacant space, it was the presence of something else — a bulk, a thickening — emanating from it, in a silent voltage.

Posters and maps were tacked on the walls of the departure lounge, intricately detailed charts of the shattered tip of South America and its clot of scattershot islands, peninsulas, fjords, and ruse-like channels which only lead to cul-de-sacs. This was Tierra del Fuego. The margins of a map of southern Patagonia presented the native animals and the indigenous peoples, the Tehuelche, before their cultures were broken by the Spanish invaders. They explained how the southern tip of Patagonia got its name, from the fires the Tehuelche lit at night to warm themselves in the frigid southern winter, and which the Portuguese navigator Magellan and his entourage spotted, guttering in the wind, from their barques.

We were inching, incrementally, away from the world. Only one flight and a turbulent strip of ocean lay between us and our destination.

"You'll see," Paul had told me at Conference. "There's no place quite like the Islands." His expression was difficult to read — not encouraging, rather stone-faced, as if the Islands were a test to be endured. I knew almost nothing of the Falklands, although I was old enough to remember static-lashed footage of the 1982 war: battleships on fire lumbering through turbulent seas, fatally wounded by torpedoes, miserable young men wearing green trudging across bleak moors.

We boarded the only commercial flight to the Islands since Argentina stopped planes flying through its airspace, the weekly LAN Chile flight. The stewardesses announced we were on our way to what they diplomatically referred to as "Las Valkan/ Malvinas." As the plane pulled away from the ground and banked toward the ocean, I felt a rush of fear; I had just stepped off a cliff. It seemed impossible that any civilization could lie beyond the black shores of Patagonia.

The sea was sundered by whitecaps. As we climbed we were buffeted by strange roaring sounds — wind shear cutting across the wings. The coast of South America curled behind us, then was swallowed by the horizon.

An hour later, out of the ocean jagged basalt reefs emerged; the sea foamed at their hems. Behind them were low mustard-coloured hills. I was glad to see it, because it was land, even though nothing in that first glimpse suggested welcome. The pilot descended and we flew two thousand feet above a landscape striped with dark cloud and rivers of stones. Light, dark, light, dark. Like a zebra.

We were told to fasten our seatbelts. *"Señores y señoras, vamos a aterrizar en las Valkan/Malvinas."*

"Did she just say we're going to terrorize the Falklands?" Tilly, the soil scientist in the seat beside me, asked.

"Atterizar means to land," I explained. We strafed the stern yellow hills of the islands. Peering at the horizon, through veils of low cloud, I made out the blue fissure of Falkland Sound broiling with whitecaps. It was more likely the islands were going to terrorize us.

We landed sideways, sheared by a gust just as the pilot sunk the wheels onto the tarmac. With the rudder he wrenched the fuselage straight in the same second the wheels thudded onto the ground so hard that Tilly grabbed my hand. The islanders clapped. The rest of us prised our fingers from the armrests.

In the baggage reclaim area, we were greeted by signs. *DANGER* said one, although it declined to elaborate. On the other side of the baggage room was a skull and crossbones; underneath it was a small spaceship device with rays of light extending from it — land mines. *ALERT STATE: BIKINI* said the sign above the luggage carousel.

"Is that bad or good?" I asked Tilly.

"They could have said *ALERT STATE: ONE PIECE SPEEDO.* That sounds more serious."

Men in fatigues hefted sausage bags onto the trolleys. We lined up in the queue for "foreigners" — non-Islanders, I supposed. Military police stamped us into the country. The eye of the man who scrutinized my British passport snagged on the place of my birth. "I spent six months there on secondment!" he declared, naming the military base near the town where I was born but have never since lived, delighted to find a random Canadian in this sink-plug part of the global basin.

Then, before we knew it we were out the door and fending off shafts of sunshine so bright and pre-emptory I had to put on my sunglasses. We were driven away from the airport to the capital, Port Stanley, nearly an hour away. An old school bus that rattled fiercely on the pitted road afforded us our first views of the landscape.

I stared again at the mustard hills streaked with rivers of stone. These tumbled between bald-pated peaks, veined out into the flatlands below, cascading a jumble of sharp-edged boulders at the edge of the gravel road. Savage-looking sheep walked among them, picking at stones. The wind slashed us. I had read that the Falklands is one of the windiest places on earth. My eyes scouted for trees in vain. Large cinder-coloured birds flew ponderously in the sky, oblong-bellied, like military bombers — skuas.

In Stanley we checked into our hotel, the Upland Goose. Pink walls, floral carpet. A vaguely musty smell. The walls were very thin, the roof made of corrugated tin.

Soon we were greeted by a man with a Midlands accent in a chef's uniform. He emerged from nowhere in a blood-stained apron, a cleaver in his hand, to give us keys. "We try to please, we try to please," he said, as if he had been put on repeat. "I've been brought down here to clean things up. Yes, it's all looking up at the Upland Goose." He grinned.

We women were billeted in a house separate from the hotel, a sort of annex. It was too far to walk with all our luggage. Caroline, a diver, had huge banana-shaped bags. It turned out these were her flippers, as well as her personal wetsuit.

"I know it's round here somewhere," the taxi driver said, "but I can't for the life of me figure out where."

We all looked at each other: how could *a taxi driver* not know where a house was in this tinpot outpost? We showed him the number: 38 Fitzroy Road. Finally he pulled up on the curb and asked a lone man walking a wolfish dog.

"I guess he didn't pass the Knowledge," Caroline whispered.

"Been here long?" I asked.

"All my life," said our grey-haired driver. "Falklands born and bred."

The house was serviceable but anonymous. There was no point in trying to redecorate or modernize it. The nearest IKEA was in Houston, 10,084 kilometres away, according to Google Maps — I checked.

I mentioned this to Tilly. "You hardly need to worry about soft furnishings. We won't be here long anyway, unless the ship is delayed." She put her hand to her mouth. "Shit. That's the Antarctic equivalent of saying Macbeth in a theatre."

Later that night we were told the ship was late arriving from South Georgia. We would cool our heels in the anonymous house for four more days.

NOVEMBER 29TH

Scalding rain. Starched wind. The names here sound made up, topsy-turvy names for the tip of the world. *Whale grass. Darker ferns. Diddle-dee.* The islands — 778 of them in total in the archipelago — might all be like this, wind-polished cirques, shattered quartzite mounts scattered with helicopter debris, with wind-preserved tubes of Argentine toothpaste left behind by the invading forces. Sea lions bark like customs officers at their perimeters. Orange-beaked gentoo penguins breed in their thousands. The islands are a rookery for beasts of the underworld.

Union Jacks, sheep everywhere, including on Falkland coat of arms. Christmas decorations strung from the ceiling, or hanging, sun-bleached, on a tree in the corner, the summer sun ripe in the sky. A smell of hot pine — pine trees heated by the sun. Also coconut — some flower? Dilapidated tin houses, wind-worn. Otherwise like a Scottish suburb, some hill-perched fishing town. Steel water threaded with kelp.

This is one of those places where animosities are intensified. Feuds, dislikes, outright hatreds — all these coagulate here. No distractions, no escape. The kind of place that makes you think, Why am I here? An instant exile.

Within a day I had walked every street in the town. Two chunky war memorials stood on either end of the settlement, freshly coated in red poppies. Even without the memorials I could sense the presence of war. This was a taciturn but rebellious land. It wouldn't let itself be taken easily, this place which was no prize of any country, a scrappy Gondwanaland archipelago orphaned from its geological parents, Patagonia and South Africa.

We FIDS, as we were now called — from the days when the British Antarctic Survey had been the Falkland Islands Directive Survey — wandered around town, looking at the money in our hands in frank disbelief, a counterfeit-looking currency adorned with sheep (and not, as I had expected, with Margaret Thatcher, the Islands' saint and saviour) we would be able to spend nowhere else.

Royal Marines marched up and down the streets in their wine-coloured caps, or drove by in an armada of old Land Rovers, little Union Jacks fluttering from their radio aerials. We found a man who made taxidermy sculptures of whales and dolphins, and visited the museum-like display in his front garden. We took photos of ourselves in the red telephone box outside the post office, to attest to the fact that we had flown all the way down the planet, some 13,000 kilometres, only to end up in the United Kingdom.

Time stood still in Stanley, in more than one way. In the pubs, people stared intently. Men ranged round the bar would turn, as one, and look over their shoulder. People had wind-reddened cheeks, blue eyes so light they looked transparent; their skin had a reddish, almost Amerindian tint. The islanders spoke with a twang that sounded vaguely Australian.

The emptiness of the place was thrilling. It was there in the ocean that broiled just beyond the thin strip of land that cosseted Stanley Harbour, the fact that there was no land between the islands and Antarctica, or on the way to Australia; in the omnipotent skies marked by strange funnel-shaped clouds and torn apart daily by the Tornado fighter jets on sortie from the base at Mount Pleasant; and in the sere landscape of the islands themselves, whose only history and culture had been these tin sheds and hardscrabble farms. There had never even been a cinema in Stanley, we learned, "or a bowling alley," an Islander said, inexplicably. Superficially it resembled an outer Hebridean town, but there was nowhere to go — no planes to fly you to Glasgow in an hour, say. Nowhere at all, except the mother country 13,000 kilometres away. By contrast the coast of the enemy, Argentina, was only 480 kilometres offshore.

In the evenings the sky turned the colour of gasoline. Breathing the air in the Islands felt like drinking champagne. I was light-headed, buoyant.

We entered a pub as a group. All eyes were fixed on us. They watched as we lobbed mistimed darts into the dartboard.

"Why do you think they stare so intently?" I asked Max.

He shrugged. "Fresh meat."

I'd come across Max many times by now — we were all serpentining in and out of each other's existences, bumping into each other in the town's single supermarket, in the swimming pool, as we passed our days as consorts-in-waiting to the Antarctic.

He was tall and blond. He had green eyes, which I usually distrusted — they don't seem entirely real to me. His was a face of drastic slopes. There were hollows under the ski runs of his cheekbones; his mouth was beautiful, although there was a cruel cant to it. He looked newly minted, sculpted into perfect life by an anatomist.

Max had a slightly militaristic air, as if he had reluctantly forgone a natural vocation to issue orders. I could tell he was one of those people who either liked you or he didn't, and he didn't think much about it. The decision made itself for him.

"You remind me of my ex-girlfriend," he said.

"Is that good or bad?"

"You have remarkably similar physicality" was his retort, his voice clinical, or perhaps that was his very faint German accent. I felt as if I'd been subjected to an examination.

In other circumstances I might have avoided Max, but there was no avoiding anyone in Antarctic World. During those days in Stanley, we all banded together. Suzanne was on the ship; she had been able to take more time away from her life than I, and had the benefit of a ship tour around the sub-Antarctic island bases — Signy and King Edward Point, plus a call at South Georgia, where Ernest Shackleton is buried.

In those days we all fell easily into each other's company. We went for meandering walks to see the penguins at Volunteer Point, to play five-a-side football. Through all this, Max and I carried on a long conversation which we picked up effortlessly after hiatuses.

On the first of those waiting lounge days in Stanley after we'd been told of the ship's delay we walked out of town to the lighthouse; its red and white plinth was a popular destination for a walk out of town, past the obligatory minefield and the dingy sheep and the jerry-built shed that housed, we were told, stacks of

automatic weapons for the day when the Argentines come again and the Islanders would queue up for them the way Londoners queue to buy sandwiches at Pret A Manger.

Max was a good talker. He was a climatologist and so interested in the larger picture, in how pieces of the puzzle fit together.

He told me he was part French and part German. He spoke both as well as English. He refused to divulge his father's profession, saying he was a "kind of financier." I thought, Arms dealer. Venture capitalist. Both, maybe. Whatever his father did for a living, Max had the caginess of the very wealthy.

Until sixth form, he had gone to school in England. This explained the lack of accent. Although it seemed that his father had also been based in New York for three years, and a vague memory of an American accent tugged at his vowels. He'd studied physics and then glaciology. He had wanted to join the Army and put his skills to work, perhaps as an engineer or situation modeller, but he'd become fascinated by science. "It holds the key to explaining everything," he said. "All the past, and all the future."

Max's PhD would investigate how the ice sheets formed. And why, once they had reached their apogee in numerous cyclotherms — cycles of freeze and melt — did they start to retreat? Why was the planet put through this cold trauma of successive ice ages — so that it could flourish anew? He told me the evidence suggested that, while the ice age killed off many species, after it was over birds, animals, insects, and plants took a gigantic leap forward. Earth flourished in cold in a way it never did in heat.

"The truth is, certain species do well out of ice age extinctions," he said. "Humans, for example. The last one cleared the way for Homo sapiens to become the dominant species."

I was so busy processing this idea — that ice was not a symbol of a frozen hiatus, that it might accelerate human development rather than hit the pause button on it for thousands of years — that I failed to hear his next question the first time.

"Will your novel be about climate change?"

"You can't really write novels based on issues. They have to be grounded in emotion, in incident and character. I want to give life to what I experience there."

The wind flung our hair into our mouths. The wind really was different in the Falklands. It arrived unannounced, like a single piece of energy, as if it had just been generated from the ground. I picked my hair out of my mouth and laughed.

Max scowled. "What's so funny?"

"I'm trying to describe being a novelist here. With you."

His scowl deepened. I could see him trying to parse whether the *with you* was a positive comment.

"It's the unlikelihood," I said. I could feel the moment — made up of the bizarre fission of me trying to describe something as urbane and effete as writing a novel while standing with a twenty-three-year-old climatologist-in-progress at a lighthouse at the end of the world while we were torn by cyclonic-strength winds — slipping away.

I let it go. A small loneliness settled inside me. Max caught it, or caught something, because he set about trying to reassure me as we walked back to town, letting loose a volley of acute questions: Do you believe in absolute truth, that such a thing exists? Why are you interested in an emotional realm? Do you write out of will or desire? What is the relationship between them? What is the relationship between intellect and emotion, in a writer?

With most people I would take this kind of questioning as a bid for supremacy, part of the constant tug-of-war for dominion that humans indulge in. Max might have been testing me, but he was also genuinely trying to understand. I was surprised by how readily, and how certainly, I answered his piercing questions. Max was energetic, curious, articulate. In those hiatus days in the islands, I'd seen through small gestures how he could be sweet and generous. But I had a sense that any sign of weakness — of intent, of character — might attract his most obvious quality, that go-for-the-jugular instinct of the confident young man. He was not a people pleaser; he did not think about decorum or the effect his questions had on me. He simply wanted to know, so he asked. There was a bracing clarity to this approach, although at times I suspected I was just another ice sheet and he was coring me, excavating for data.

We might as well have been separate species, Max and I. He was young, handsome, confident; he had been raised on a diet of certainties and accomplishments. I was fourteen years his senior — long enough on the planet to understand that life can be a kind of dismantling in which you are taken further and further away from any original certainties and power you might have possessed, further from yourself, until you are only a moon orbiting the sun of who you once thought you would be.

DECEMBER 1ST

Morning in Stanley. The wind is beginning to pick up, at eight thirty a.m. Soon it is as stiff as a plank. Clouds are coming in from the west. The coconut smell turns out to be yellow gorse.

The *QE II* is expected in today. We can see the ship's motorboats coming ashore. The tourists get in our way. They buy stuffed penguins in the gift shop and yogourt in the West Store. The passengers are dapper men with cream-coloured scarves wrapped around their necks and thin women wearing bronze-coloured ankle boots. "Argies," the West Store cashier says darkly.

Today we are going on a planet walk — this is one of the many quirks of this place, that around Stanley Harbour is scattered a model of our solar system. The planets are placed at intervals which are to scale with the sun. The sun itself is located front of a shipwreck, the *Jhelum*, but Pluto is out of town, apparently, somewhere on Mount Tumbledown.

Now the weather is changing. A rain squall has just arrived, emitting fierce wintry hail, while one side of the sky remains placid and vernal. "That's the Falklands for you," someone says. "Four seasons in one day."

The outlandishness of the Falklands played its part in those hiatus days. We were all decoupled from our usual lives, and drunk on the effervescent air. I walked the streets of Stanley grinning, my sunglasses deflecting the sun's near ozoneless orb, past bad-boy bar Deano's, the bakery that churned out sub-Greggs white rolls, the 1970s constabulary, the teal-headed upland geese snoozing in the sun or efficiently mowing the town's greenery, aware that I had freed myself from a nameless prison by the simple fact of coming to the end of the world.

The next day we went for an outing, piling into Land Rovers. We were driven over bare hills that turned flaxen in the sun. We pulled up at a trim white farmhouse overlooking a crescent-shaped beach and set out walking over peat bogs, the peat like wet espresso. Upland geese squawked at our approach. All the shrubs had been battered to bonsai by the wind.

Ahead, the penguin rookery awaited. We heard an unfamiliar chatter, a tight sonar tapestry of honks. Penguins were talkative, as it turned out. Behind the dunes they stood beak to beak and upright, like mini statesmen, having earnest conversations with each other. On this beach, thousands of king penguins congregated to rear their young. The late November sun had a rapier quality I'd never felt before. In its gold light, the ocean was a sweep of blue sequins.

Max and I walked away from the colony and started down the empty white sands fringed with jade shallows. We watched as small black forms emerged from the waves, slick and glittering in the sun. This was my first experience of penguins in the wild, and they seemed from a different dimension. Here they were, emerging from the ocean and shaking themselves dry in their neat grey and black wetsuits, only to be sandblasted on the beach. They paused on the tide line, shut their eyes against the sandstorm, and soldiered toward their burrows in the dunes.

Max and I stood on the beach and picked grains of sand from our eyes, transfixed by how different, how new, the world could feel, just by changing one's position on the planet. It was pleasing to be so disrupted from reality.

Max and I were commanded to move on — under no circumstances should we disturb the penguins with our presence. Others joined, and the tallest of us, including Max, positioned themselves against the breeze as windbreaks. Max towered above everyone, his head tilted down to better hear what people were saying; the wind ripped words from our mouths. From time to time, he would throw his head back in laughter or duck down closer to his interlocutor and laugh. He had an athletic, martinet stride which didn't quite seem to belong to this age.

On our last night in Stanley before boarding the ship, the Women's House at 38 Fitzroy Road hosted a farewell to the Falklands dinner. Nils came, as did Max, Emilia, Patricia the soil scientist, Tilly, Veronique the "squid woman," who was throwing over her usual specialism to spend the summer tagging penguins on a sub-Antarctic island, Caroline the diver, and me. Bottles of Savanna cider and Chilean wine were piled on the table.

Max walked in the door. "Can I see your books?" He said it without introduction or hesitation, as if he had been thinking about this question for some time.

I showed them to him. He read a page or two. In front of everyone he read a line out loud — I forget which one — and said, "So does that come from your writer's instinct to feel the pain?"

The question hit me like a flare. "Maybe there's another instinct you've missed."

He grimaced and looked away.

The party continued. Max had brought a cake. When it was ready to be served, he turned out the lights and lit candles on the table.

I had brought an issue of the *London Review of Books* with me to read (thinking, correctly as it turned out, these would not be thick on the ground in the Antarctic). In it was a review of an Edvard Munch — he of *The Scream* — exhibition.

Late in his career Munch wrote, "The second half of my life has been a battle just to keep myself upright. My path has led me along the edge of a precipice, a bottomless pit. . . . From time to time I've tried to get away from the path, thrown myself into the throng of life among people. But every time I have had to go back to the path along the cliff top."

Munch's topics were depression, despair, the pain of jealousy. He became a painter of the inner life, particularly his years among the bohemia of Berlin and Oslo (then called Kristiania).

The painting that accompanied the *LRB* article was a study for what would eventually become *The Wedding of the Bohemian*. I looked around our Falklands table: a group of people together for a celebratory dinner, the silver light of the far south, pizza crusts, empty bottles of beer and Chilean wine, a demolished cake. In Kristiania, Munch's guests, their visages effaced, a Spartan table, two bottles of wine, a casserole, and a dessert. The happy couple in their modest wedding attire sit at the far end of the table.

Munch was a melancholic romantic, a handsome man who never found his match. He was always attending some dinner party or other where a beautiful, desirable woman said something cutting. Despite his good looks, Munch did not attract love, or luck, and he knew it.

The Falklands' sudden spring, the gorse with its Bain de Soleil scent, the unfamiliar sharpness of the sun and wind in the islands had revived me. But now I felt like the disconsolate woman at the dinner table in Munch's painting, stripped of joy; in its place were stark white stripes of anxiety. In Max's comment I heard the sound of knives sharpening. For me, criticism and intimacy had always run on the same current.

Night lingered on our forgotten latitude. A brawl had broken out inside me. I have always been prone to sudden gusts of depression; my mood can turn on a dime. I felt like flinging wineglasses across the table. The moment when Max might have been a friend slid by, observed only by me, on its way to an unknown destination.

Shades of Light closes at six. By seven I am out; in between is the till reconciliation and the daily dust-down — dust accumulates in thin pelts settling on the stained-glass butterflies, hand-dyed silk scarves, and artisan greeting cards, the dried starfish and sand dollars. Thick amber sunbeams pour into the store. It is nearly spring; winter is finally relinquishing its grip on the land.

The walk home is ten blocks by the city's well-ordered streets, but there is a shortcut along the old railway line that I take in the hours of daylight. The town is changing: the old flour mill and the Bata shoe factory are being turned into "luxury apartments." Billboards show white spaces razed of walls and appliances made of brushed chrome.

It has been forty years since passenger trains stopped in town. Now there is only the grain train, which slides through town twice a week to load wheat and rye at the silo. An old warehouse next to the track has recently been converted into a laundrette. In the winter I walk through clouds of steam emanating from its windows. Now, in early spring, woolly soap smell floats from the windows.

I insert a cassette into my Walkman — Siouxsie and the Banshees. I do not hear the car pull up.

"Hey . . . HEY!"

I tug the earphone out. The blue-white car has caught up with me just before the train tracks.

"Where ya gon?" Michael does not have his patrol light on.

"Are you following me around?"

Michael grins. "Just keepin an eye out. It's my job, remember?"

"Where's Donna?"

"Home, doing homework. Shouldn't you be too?"

"Some of us have to work."

Michael drew his head back into the car. "Just be careful.

Keep your eyes out. I wouldn't wear the Walkman, not with what's happening. See you around." He says it *seeyer*.

"See yer." I wonder if he has noticed I am mimicking him. But he just drives away.

In spring the woods are mulchy and secretive. Since I was a child, I have always hallucinated in forests — I see creatures, dark-pelted, humanoid, melting between the trees, or giant wolf-bears. I am never frightened by these apparitions. They seem to want to avoid me. These images knit together across my retinas, shadowless.

Donna and I walk through the paths of the woodlot by the dam. The trees are bones against the sky. In another week they will have leaves. Spring hits like a hammer in this province.

Geography is my best subject that final year of high school. We learn about the Carboniferous Age, when the coal that once brought riches to the province was forged. This was a world of ferns, bog, and immense pressures as fallen soggy trees were compacted into graphite. This, we learn, is the same substance as diamond, essentially — only a single bond in their structure separates them.

Mr. McIsaac, our geography teacher, tells us to look for the land's memory of the ice sheets which once covered the province, to read the history of its abandonment by ice: its chapters of boulders, its lonely paragraphs of drumlins, and the furrows of mica and malachite. This is the tumble left in the wake of a hasty retreat as the ice pulled back toward the north, a giant snake recoiling. He tells us our land is the product of a vanished sheet of ice that had clawed its way almost down to the Mediterranean.

Donna and I drive out to the woodlot to walk, talk, and

occasionally to sit on a log and drink a beer in quick fugitive gulps, secure from the gaze of her parents, jiggling our legs to keep ourselves warm.

I am seventeen; Donna is a year older. I have skipped a grade and have been put in the accelerated programme, a term which always makes me feel like I am being shot forward into the future in a catapult. We will graduate from high school in eight weeks' time. After that, for some of us our futures are unmapped. Even those of us headed for university don't really know what to expect. We don't know how fleeting it will be, this hiatus in our lives; for the moment our lives are claimed neither by the past nor the future.

We pad along on a carpet of pine needles mashed into a paste by the winter weight of two feet of snow and ice. Clumps of ashen mushrooms bloom in dark pools where no light has penetrated all winter. Layers of defrosted maple leaves sigh underneath our feet.

We find a log and pull two bottles of beer from a sturdy paper bag. We hide the bottles behind us. We have to keep an eye out for park staff or worse, police; drinking alcohol in a public place is illegal. We fear the police less than Donna's parents, particularly her monumental, stern father, a local concrete magnate and strict teetotaller. Even Michael, Donna's older brother who is on the force with the RCMP, might not be able to save our necks if we are caught.

"Michael pulled me up yesterday."

"Lucky you."

"I think he was trying to tell me not to take the shortcut through the railway yard."

Donna takes another swig. "Well, maybe you shouldn't."

"I don't want to live in fear."

"Nobody does. It's just a practicality. Given the situation."

"Is Michael looking for him?"

"Everybody's looking for him. Michael won't let me walk home. That's why I get picked up from school in a patrol car every day. Needless to say my popularity has taken a nosedive." Donna hits herself on the head, very lightly, with the bottle.

"You'll always be popular, don't worry."

I have only three or four friends, a legacy perhaps of having moved to this city in junior high, too late to establish myself in the local hierarchies. My other friends — Christine, Summer, Morgana — are children of hippies. They live in homes with log-cabin interiors and kitchens stocked with lentils. Donna is my lone normal friend, from a long-established local family. Her religious parents and six siblings keep Donna under constant surveillance. A particular thorn in Donna's side is the watchful, restless eye of her older brother Michael. He used to be an ally until he joined the RCMP. "He's on his high horse now," she says. If he discovers anything of her real life, she will be carted off to Pentecostal camp in Ontario for the whole summer.

The climate is completely different again from the island where I was born, ten hours away by road on the Trans-Canada Highway. This town is a place of rigid winters and cloying, humid summers.

The edges of the air are cold. We shiver in our down jackets as we walk the narrow paths lined on either side by thick spruce. This province is so densely matted with trees that some of its landmass is still not properly charted. From a plane you see its black tangle threaded by the sand-coloured arteries of woodlot roads. The town where we live is large in terms of its population, but it feels like a small town, with its wide eventless streets and highway strip malls, its hockey rink flanked by a Kentucky Fried Chicken and a Dairy Queen. I moved to it to live with my mother, who I only met some six or seven years ago, in order to finish high school. Before that I lived with my grandmother in a provincial capital city where I took art and tap dancing lessons and had classmates from Colombia and Germany.

"I can't wait to get out of here," Donna breathes, casting her gaze upward, past the branches which razor the sky. "The next two months are going to feel like forever. You know, I don't even know what these trees are called."

Sumac. Larch. I am learning to identify the local trees in geography: Siberian and European elm, hawthorn, laurel, jack pine, trembling aspen, tamarack — trees that bequeath their names to the streets in this town, along with its scant human history, the two or three names repeated over and over again, the upright, English names of the early settler barons of the province who had immediately set about flensing it of its trees, minerals, fish, and animals.

"We shouldn't stay here too long." Donna shivers. "Dad says that when it's this cold so late into spring you can expect a hot summer. But he used a word I didn't know," she frowns. "It started with a T. Like torrent."

"Torrid," I say. "A torrid summer."

4. A VORTEX

frost flowers
Growth of ice crystals by condensation from the atmosphere at points on the surface of young ice.

DECEMBER 4TH

We set sail into the Southern Ocean — the officers still say this, "set sail," although it is the ship's four engines and a type of diesel called marine gas oil which propel us through the water.

The ship reverses out of the jetty and twirls on its rear stabilizers. We glide out of the placid waters of Stanley Sound, a narrow gut of water separated from the open ocean by only a kilometre of land. The calm is deceptive; on the other side of the headland lies a different sea. "We're heading out straight into Force 8," Captain David warns us. "You'll all have to find your sea legs fast."

As the ship begins to grapple with the swell, we play Scrabble in the bar. I watch my opponent Nils quickly turn lime green. He abandons the Scrabble board and makes a hasty exit. The worse the motion, the better I feel. The sudden unsteadiness of the ground beneath our feet invigorates me. I do a turn as barmaid, planting my legs against the sink so as not to be flung face-first into the bar's ice-cube-making machine.

Next on the agenda are safety drills. We sit in the bar, accompanied by life jackets and smothering glassine shields called smoke hoods which turn the world amber. Mike the purser instructs us in doom scenarios in his Glaswegian buzz, so thick I pick up only every third word: "*Brrrr* . . . iceberg! . . . *Grrrrr* . . . fire drill! . . . Muster point . . . *brrgggrrr* . . . all of yez!"

As it turns out, fire is as much of a danger as being overwhelmed by a freak wave, or engine failure. Fires on board can rapidly become uncontainable. Once started they quickly rage out of control in the confines of an oxygen-rich, flammable environment. We practice fire-fighting, donning smoke hoods, our sucking breaths loud in our ears. On deck we learn how to manhandle the thick snakes of water hoses.

Safety training over, we sink into our ship's cosseting routine: two seatings for dinner (we are too many to fit in the saloon all at once), one at six, one at six thirty, lunch is at twelve and twelve thirty, and breakfast at seven and seven thirty a.m. Our names are assigned to slots: we cannot wander in when we like. We have fabric napkins with silver napkin holders, and on these our surnames are pasted. Gash — cleaning duty — is also assigned.

At night in my cabin, I rock in my sleep. I dream of tsunami-like waves that capsize the ship, although everyone seems to survive, and the event takes place in tropical rather than Antarctic waters. Ejected from the ship, we try to hang onto the hull but our fingers slip on gooey seaweed. I wake after these dreams and register a heightened version of the emotions I feel as I move around the ship by day, or plant myself in front of the giant windows of the bridge — isolation, insecurity, awe, gratitude.

There's another element to being on board a ship which is hard to define: a claustrophobia twinned with excitement. No new input will come into our world for the next two weeks. We are a sealed society. Over the long term, this can settle into the boredom and inevitability that breeds discord. But in these early days, our confinement generates an emotional mystery, a sense of drama. Because there is no escape from each other, a reckoning might take place.

Max hovered on the threshold to my cabin. I was busy tacking up postcards of Frank Hurley's photographs of Shackleton's *Endurance* expedition. Men with euphoric faces beamed at me while drinking tea out of billycans and ruffling the fur of one of their sledge dogs.

"What will you do?"

"What do you mean?"

"All day, while we're doing science?"

Everyone on the ship had an assigned role: the officers on their clockwork watches, four hours on, four off; the scientists doing twenty-four-hour oceanography; the cooks and the stewards making Yorkshire pudding in the kitchen.

"I'll write."

Max leaned his shoulder against the lintel. "Write what? Nothing's happening yet."

"I'll write about things not happening, then."

He scowled. He knew I was humouring him and was unhappy about it.

I relented. "I'll write poetry. I'm going to write a piece called 'Night Messages.' I got the idea from the captain's night orders book on the bridge."

Night orders is what the captain writes before going to bed, if he is not on watch. They are nautical instructions, about courses made true, knots, hazards to look out for. They all end with the same phrase: *Call me if required, or if in any doubt whatsoever.*

We had been at sea for twenty-four hours. I felt like sleeping all the time. Apparently this was a well-known effect of the beginning of any journey by ship; in heavy seas especially, the inner ear has to work hard to maintain balance in a suddenly moving and tilting world, and this tires the brain. I was also getting used to the constant electric shocks — electrical charges build up in the enclosed environment — and to staggering along the corridors, bouncing from one wall to the other, or keeping a leery eye on the lip of the soup bowl at lunch lest it slosh over, and a thousand other tiny adjustments to life on a roller coaster.

There was an immediate ease to ship life too. I sensed how the purposefulness of being at sea was its own justification. Our entire reason for being was to move across space. We were on our own, ploughing through the worst seas in the world, on an important mission. For the first time in my life, just to be in motion felt thrilling. But at the same time there was a hesitant, uncertain timbre to life on the ship, like a faint music. We were on an ice-strengthened ship in the Drake Passage, widely acknowledged as the most challenging seas on the planet, and every hour at sea felt like a fresh peeling away from the world. Events and families as much as the background trivia of life — newspapers, train journeys, TV, shopping, espresso — began to take on an outlandish glint.

The Antarctic world was unordinary — this had been clear from the beginning, at Conference. Now, a power hierarchy and social structure made itself plain to us during those first days on the ship. Our first inkling was on the afternoon we were due to leave the Falklands, when a strange thing happened.

The captain was conducting a briefing in the bar to inform us passengers what to expect on the passage to Base R. He stood in front of us in the saloon, tall and slightly glacial in that way of so many Englishmen. He wore a white shirt with four gold bars on his epaulettes and brogues shined to liquid tar. His hair was thinning, but he was vigorous. He looked about fifty-five. We discovered, during the voyage, that he was ten years older. The source of his command seemed integrated into every fibre of his being. In the narrow corridors of the ship, you could feel him coming before he rounded the corner: his authority sent out an advance force field. The very air stood up straighter in deference. There was a note of ownership in all his gestures. This was *his* ship.

So we were all surprised to discover we were invited to call the captain by his first name. "Normally, if we steamed straight, it would take us four to five days," Captain David told us. But oceanographic studies on the Drake Passage water column on the journey down meant that the ship had to be stopped for periods of time. While it was not normally advisable for a ship to be still in rough seas, the *James Clark Ross* could remain balanced thanks to a system called dynamic positioning. After the oceanographic studies had been completed, we would bring cargo and equipment to two bases on the way down. "We're looking at a week and a half, maybe more," the captain said.

A carpenter who was part of our group flight from the UK got up during the captain's address, walked to the bar, and extracted a beer from the fridge. My eyes followed him. This carpenter had started drinking on the plane from London to Madrid and carried on through our stopover in Santiago, and all the way to the Falklands.

Later that day, we were informed by the purser that the carpenter was already on his way back to the UK. He had been called into the captain's office and summarily dismissed from the ship for drinking during the briefing.

"Does that happen often," I asked Mike the purser, "that someone is sent back?"

"Quite rarely. The bosses at BAS get it right with who they send down here, most of the time. But it's better and cheaper to do it now. Think about it," he said. "If the shit hits the fan — if something goes wrong with the ship, say — do you want to have to shoehorn a drunk guy out of his bunk and into a life raft? And once he gets to base there's a £10,000 ticket to fly him home."

We all had prices on our heads here, I was reminded. It had cost a certain sum of public money to put me into the Antarctic. During our launching into the seas in the lifeboat drill the day before, the plunging drop into the ocean as the winch let us go and we crashed onto the waves, I felt a scrawl of anxiety in my stomach. I had a job to do here. I hoped I was up to it.

In those first days of our passage south, the weather was calm, even balmy. I took to spending hours outside on deck in the last supra-zero temperatures I would feel for several months.

I hadn't bargained on the bird life of the Southern Ocean. We had a voluble, careening escort of storm petrels; cape petrels; sooty, black-browed, and grey-headed albatross; and small cessnas which turned out to be the wandering albatross. All capable, fierce-looking birds, they rode the troughs of the waves in an oracular manner. They knew what the sea would do before it did. The sea spasmed, then flatlined, washed itself in the wind, tossing spray, then funnelled, slapped back by the ship, and still the birds flew with their stomachs only a millimetre from its surface. I watched this pas de deux with the waves, fascinated by their precision. They didn't even get their wing tips wet.

The grey-headed albatross was the most elegant, I thought, with its ermine head and nape and intelligent, roving eye. I watched the wanderers from the outside deck, where they skimmed the waves in the wake of the ship. You can identify Coleridge's albatross immediately by their size — they have a total wingspan of over two metres — and by their phosphorus-white colour. Wanderers also have a large, flesh-coloured bill which gives them something of the thuggish look of the skua. They fly as if motionless, their wings held static like aircraft. I watched a single bird for an hour, thinking, Now he must flap his wings, now, now, now. But he never did.

Life was slowing down. My bird observation installed an unfamiliar quietude. It had been a long time since I had looked at the detail of the land or the sea. For thirty years I had been living a headlong life, soaked in stimulus. I would have to re-train myself to notice detail again, to notice the innocuous. One of the deprivations of so-called modern consciousness is a latent expectation that observable reality should deliver us a personal satisfaction: that every incandescent sunset or every mild morning is blamelessly complicit with our desires. It is an egoistic fantasy, of course, but it is remarkable how difficult it is to not expect the next jolt of satisfaction around every corner. In environments barren of stimulus one can encounter an interior horizon you didn't know existed, a kind of calm, interior core of consciousness, but this experience is hard to engineer in cities.

It was useful that we were going there on a ship, I thought, instead of in one of the BAS planes which flew people into the Antarctic from the Falklands in four hours. On a ship you approach the Antarctic gradually, crossing a series of thresholds. The world outside your cabin porthole becomes more and more unfamiliar, and you enter into the alternate reality of the continent, a transition signalled by a new language — *ablation, sublimation, drift, terrane* — never mind the Stygian mists and chrome skies. The words of the French glaciologist Mathieu returned to me: "The Antarctic looks like nowhere else. At no other point on the planet can you see those colours, or that relationship between land and sky."

But for the moment we were still in the known world. The sea looked lumpen and agitated but not lethal. It was cold, but no colder than England in the winter. In the bar those first nights at sea we played Trivial Pursuit — raucous, winnerless games. We played blackjack and snap, we laughed. We did pub quizzes, listened to each other's CDs. We drank and talked in aimless wide ellipses — purposeless, floating discussions of the kind I hadn't had since my undergraduate degree.

At night I retired to my cabin, but I was not alone. I had a chaperone. For the whole trip, he, or versions of him, would accompany me. One day a black-browed albatross, the next the giant of the southern skies, the gaunt white wanderer. He flew level with my porthole, day and night. From time to time, he turned and looked into my cabin porthole and I locked eyes with him. His gaze was steely, disapproving — an envoy sent to gather intelligence but not friends.

DECEMBER 7TH

Max wanders the ship at night long after I have gone to bed, in a self-appointed mission to map those spaces listed on the internal telephone card in each cabin and which we wonder about: the Rough Workshop, the Cool Specimen Room, the Transducer Space.

"Shouldn't you be sleeping?" I say. After all he's on duty for the oceanographic studies for twelve hours every day.

"I think sleeping is a waste of time."

Max's statement bears his trademark certainty. He has never thought about this, he just decided it. What do I do, in contrast? Ruminate and examine. Procrastinate. Plump for fairness. I have become one of those diplomats of existence, trying hard not to upset obvious truths, treating them like ageing monarchs with an urge to sharpen swords. Meanwhile the tyrants, neglected, accumulate.

Sleeping *is* a waste of time. Why do we do it? Everything Max does or says — picking up his dinner knife, pouring himself a beer — carries a low shrill sound, like an unfamiliar bird's call. It says, I am alive. Too alive to submit to that nightly coma we call sleep.

On the ship we all saw each other constantly; we met by chance or design twenty times a day. We were coming into focus for each other at an accelerated speed — faster certainly than in London, where it can take three weeks to secure an appointment with your most treasured friend. Our confinement made it difficult to distinguish between accidental and sought interactions. There was a physical code on the ship for signalling you were available to talk: a closed cabin door meant *I'm busy or sleeping*. A cabin door with a curtain drawn over meant *I'm available but working*. An open cabin door was an invitation to come straight in.

Max poked his head around my cabin curtain. "How's it going? Finished that novel yet?"

My scowl, he knew by now, was an invitation to talk. He was on a break from reviewing his supervisor's ice stream data in the lab. We see-sawed with the ship's motion as he told me about his life. School was still close enough to his experience to want to talk about it. He went to "prep school." The term was foreign; it belonged to a private school system, I guessed. I had the sense not to ask what it meant.

He had gone to school in a place called Zug. He spoke of white forests, ice on the lake. His school was so far up a mountain he had to take a tram, a kind of ski lift, to get there. "I think it might be the most high-altitude school in Europe, possibly in the world," he said, with his detached, clinical air. There, he and his fellow students took school trips to ski, to see the Hellenic ruins of Crete, to kitesurf on the Kenyan coast. I envisaged Thomas Mann's *The Magic Mountain*, a sanatorium turned into a school for the fabulously privileged, their watercolour easels staking out glacial lakes.

"Who were your friends, there?" I asked.

"Russians, mostly."

"Why Russians?"

"They have a rawness that I liked. Plus they were always flying home on the weekends. They weren't around enough to get tired of them."

"It sounds international, your school."

"Oh yes. People are from everywhere — London, Malaysia, Peru, Corsica, you name it."

"Corsica?"

He shrugged. "Like I said, from everywhere. We had three daughters of African heads of state. They came to school with bodyguards. Man, those guys were bored. They were always on the phone to their three wives in the Congo. The heavies were all for show anyway. I guess no one told them they were living in Switzerland."

Max did not leave conversations conventionally. He always leapt up, as if something had just occurred to him, and departed without a backward glance. Perhaps it was an adaptation for survival among the children of oligarchs.

I embarked on my own night missions. I sat in my cabin at one or two in the morning, watching the silver-tipped waves. How strange to feel alive again, and so suddenly, revived by hardship into a world of convergences, invisible boundaries where one world is left behind and another, unknown version embraced. At the same time I felt bested by something, or someone, outstripped already by a growing and familiar duet of uncertainty and fear. But fear of what? The ship was robust. It would not let us down.

I was used to spending long periods on my own, thinking and writing. I had perhaps lost the knack for relating easily to people — if I'd ever had it — or at least for talking in a random and objectiveless manner, instead of the competitive conversations of intellectuals.

Those sabre-sun days in the Falklands, then setting sail, learning to put out fires and how not to die of immersion shock, the novelty of constant company — all this stimulus had tilted me away from my normally ruminative self. But once accustomed to this vital new world my usual doubt had returned, like a hunched cloud on the horizon.

The ship's motion had an effect on all our emotional states, too, with its rhythmic chaw through the sea, so much to me like the rhythm of thought. The sea fog came accompanied by a familiar voice. It was my internal chaperone, the keeper of shames. It spoke to me in the second person, the *you*, with its admonitory glint: How are you going to write a novel set in a place with no human history? How are you going to write something truly original, when better writers have failed? Why are you alone, at your age? What are you doing here?

DECEMBER 8TH

Four days out at sea. Stuart the deck engineer takes us on a tour of the ship. We encounter his fellow engineers in the duty mess in their grease-stained white overalls, a filthy gang of droogs making amiable cups of tea.

We move on to the labs. Here, giant tubs hold specimens dredged from the deep ocean — fish, octopus, squid, salps — a type of plankton which, Stuart tells us, is the rocket of the ocean, jet-propelling itself through the water. Clocks show GMT and local time. Urgent green signs lead us to our muster points.

Stuart is boyish; he might be forty, but his face is unlined, his eyes wide with marvel.

"You love it here, don't you?" I say.

"You know, it's strange. I've done seventeen seasons south, but each time I get excited. Each time it's like I've never seen it before."

Stuart and I are on the bridge, flanked by two able seamen with binoculars soldered to their eyes. Icebergs have begun to appear, even though we are still north of the Antarctic convergence. Those in advanced states of melt take on the pyramidal appearance of a sailboat with its spinnaker up.

Stuart and I squint into the horizon. Inside the grey light, a blue throbbing appears. We lean forward into its radiance. Is it sun filtered through clouds? It looks like sunlight reflected on pewter. The Antarctic is ripe with strange visual effects. Whatever it is, it shines with the panicked diligence of a substance about to be consumed by a vortex.

We were still seven days away from the continent, thanks to our stop-start oceanography studies. Yet we could feel its fortitude, even at that distance, its rebellion and its bulk. It makes sense that cartographers and navigators intuited its presence thousands of years before humans ever set eyes on it.

From the beginning, the polar regions have been central to the concept of climate. The word *climate* itself and the conception of changing zones of prevailing natural patterns comes from the Greeks. The word *klima* was first recorded by Parmenides, a disciple of Pythagoras. In the sixth century B.C. he posited that the earth had five zones. Even though it was not then known that the earth was a sphere, he assigned these zones to latitudes, dependent on the angle of the sun's rays on the planetary surface. Parmenides thought that the *klima* to the north was cold, and to the south it was hot — hotter than the eastern Mediterranean. The notion of climate zones dependent on latitude is too rigid to reflect the actual reality; we now know climate reflects more than just temperature, and that patterns and tendencies of atmospheric conditions, topography, and weather work together to create climate.

Around 330 B.C. Aristotle speculated in *Meteorology* that the earth was a sphere, divided into northern and southern zones, each identical, but opposite. There, he suggested, a land of cold might lurk in the furthest reaches of the *oikoumene*, the known world, which at the time extended no further than the shores of the Mediterranean Sea.

Something had to be there, the cartographers believed; it did not seem possible that the world could be ballasted by empty ocean. But how many of these cartographers would have believed that what was actually in residence at the bottom of the planet was a continent-sized sarcophagus of ice, impervious to almost all life forms and hostile to habitation? On Hellenic maps it was drawn in, sometimes with painstakingly rendered coastlines dating back as far as Ptolemy, whose "*terra australis incognita*" appeared on world maps for the next 1,700 years, before being replaced by the land first sighted by James Cook and others in the nineteenth century.

The Greeks had achieved a remarkable feat of topographical observation and imagination. That the fiery realms at the equator were countermanded by frigid realms at the axles of the spinning globe was more or less correct. Until the nineteenth century, the polar regions remained frozen possible apocalypses, unmapped and feared.

The word apocalypse comes from the Greek *apokalupsis*, a derivative of the verb to uncover, to reveal. The apocalypse is not only God's judgment, the sudden annihilation of everything, but also a revelation. In Coleridge's *The Rime of the Ancient Mariner*, the Antarctic destroys his maps, his instruments — the ways of understanding and moving through space and time — but it also unmasks the true spiritual nature of life.

As in Coleridge's seminal poem, *terra australis incognita* was a blank page which excited the imagination, but was also the perfect surface on which to project humanity's phobias and terrors. *The Hereford Map*, drawn by Richard of Haldingham and Lafford in the 1280s and which still hangs in Hereford Cathedral, shows eleven types of Antarctic monsters lurking at the southern tip of the world. Among these are a one-legged species which hold their feet above their heads, like an umbrella; men with the heads of dogs; and a species that has such a small mouth it must suck food through a reed. Finally, there are the Philli, who tested the chastity of their wives by exposing babies to snakes. The Philli thought this an accurate DNA test, believing that the legitimate offspring would remain unbitten while the bastards died from snakebite.

The Antarctic was a lair of monsters, but also of magic. I noticed how many words in the ship's officers' and able seamen's vocabulary point to a place where reality and unreality intertwine, where weather and terrain might be indistinguishable: *ice blink*, the reflection of distant pack ice on the sky; *sun dogs*, or parhelia, multiple suns, radiating around the one true sun, refracted by ice crystals. I could see for myself out of the *JCR*'s windows how we were entering a land of mirages and refractions, of whiteouts and vanishing horizons.

Vision is confounded by the Antarctic landmass and its seas, and this is one of the many reasons why the continent is so dangerous for humans. Our vision anchors us in reality; our eyes seek lines, definitions, perimeters, contrast, separations, in order to judge whether a surface will permit us to engage with it. When these are absent we are so disoriented we begin to hallucinate them. On a visual level we imagine into being what we need to see in order to orient ourselves in the world, just as our consciousnesses rely on a similar wishful manufacture to guide us through the unruly reality of our lives.

What to make of our destination, this void masquerading as a continent, a place endowed with such powerful energy that it was able to map the idea of itself onto the greatest cartographers and thinkers of our civilization before it was discovered? It is still less a destination than an idea. We will dock at no town or city; there will be no shops, cafés, offices, comfortable dwellings. Standing on the bridge with Stuart, I can already feel how helpless we will be in that *nullius*, that place where absence and emptiness rule. It is a zone of poetic force, a seemingly colourless place, yet home to the most · alluring chromatic phenomena on the planet. The only place in the world that is nobody's country. A utopia, an apocryphal vision, a conundrum. A hoax.

My shifts at Shades of Light begin just after school lets out at three thirty and end at seven, when the store shuts.

Before Elin bought the house and turned it into a store, it was someone's home. Long windows leer onto a purple-lit garden. Every inch of what would have been the living room is covered in objects for sale: rows of soapstone seals from Labrador, homemade tapers joined by a single wick which must be severed, rotating carousels festooned with earrings and necklaces and greeting cards. In the window stained-glass ornaments and crystals broadcast prisms. I have bought a few of these for myself; they now dangle in front of my bedroom window.

"Don't spend all your pay on these objects," my mother has warned me. "Or there'll be nothing left for school." She uses the word *school* but in fact it is university I will be paying for in the fall, thousands of dollars. The prospect of the money I have saved from three jobs leaving my bank account makes my head swim. I have a year's money saved, but it is a four-year-long degree.

My high school exams are just over a month away, and then, if I do well, I will be free. I feel the weight of it pressing down upon me, my future. I have been waiting for five years to be free of this place, since I came to live with my mother and her husband. No five years in my life will ever feel longer.

My mother and her husband have two babies, both boys. I try to close the screen door to my mother's house quietly, so as not to wake the children. Its squeak gives me away.

"Where have you been?"

"At work."

The house my mother and her husband, Mark, have bought is typical of this town, a wide-girthed clapboard house with a bay window and green shutters. It has an actual white picket fence. There is something vaguely shaming about these houses, with their stern windows and wooden eyebrows — those shutters again — their backyards waiting to be pummelled by children, their clotheslines and garages. Family houses, they have no other function. Unstocked by children they become mausoleums in the long blue hours of summer, the permafrost of winter.

"Come up here," my mother commands. "I have something to tell you."

I fight my way upstairs through unstable piles of baby detritus on the steps.

"Why do I have to come upstairs?"

"I don't want to be overheard."

I reach the threshold of the bedroom my mother shares with her husband. She speaks hastily, an unfamiliar sheen to her words.

"Your father — he's here."

"Who are you talking about?"

"Your *father*. Are you dense?"

I do not say, as some seventeen-year-old girls might, *My real father?* My voice high and airy with fascination. Instead my voice sounds rusted, old. "Why are you telling me this?"

"Because he might want to see you."

"Why would he want to see me?"

"Because he's your *father*." My mother says *faaather*, with the elongated "a" so characteristic of our native island and which makes our speech sound Irish. We have both changed our accents to better fit in, but it emerges in certain words, when we are angry or under stress.

"I've got exams in a month."

"I know. I didn't want to tell you. I didn't want to interrupt your concentration."

I wonder if this can be true. My mother never asks about school. She does not help me write a personal statement for university applications, as my friends' mothers do. She does not sign me up for correspondence exam cram courses. I could be planning to work in a bank, for all she knows. My mother doesn't know I have applied to the top three universities in the country.

"I've told him he can't come to the house. It's not easy for me either, you know. I told Mark he was dead."

"You what?"

"As God is my witness. I lied." My mother thumps her breast. She is given to these gestures. She has learned them in her church group; they do a lot of thumping of the head and heart there.

My mother thrusts a piece of paper in her hand. "Call him, would you? Otherwise he might show up here and create a mess."

"A mess for you, you mean."

"Just do it. He's your father, after all."

I already have a father. Perhaps my mother doesn't see it this way. But her parents brought me up — in a way her father is my father, too. I have grown up, like my mother before me, ten hours away by road in the adjacent province, on an island within an island.

But at twelve I had to leave my grandparents and come to live with my mother, whom I met when I was ten years old, and who remains a stranger to me. She and her husband live in this prosperous, sinister town next to a swollen river which floods each year, submerging the riverside dowager mansions.

These days I don't see my real father, my grandfather, much. He went off the radar for a couple of years, and has only recently sidled back into my vision. Even so, he is a diminishing blip, fading further and further out of view.

5. THE NINTH WAVE

breccia
Ice of different stages of development frozen together.

DECEMBER 9TH

Time closes like a vault behind us. The time before I boarded the *JCR* has been erased. I already can't remember who I was, what I wanted, then.

The motion of the ship is a pleasant, grinding procession. We are doing eleven knots. If she pulls out all the stops, Captain David tells me, the *JCR* can do sixteen knots — roughly the speed a moderate cyclist manages, or a cart and horse. We are bicycling to the Antarctic.

Suddenly there is a new sting in the air. We are passing over, or through, the Antarctic Convergence, a ribbon of cold water that encircles the entire continent. The water temperature readout on the bridge shows a five-degree drop over ten minutes.

I seek out Nils to explain the Convergence to me. I find him in the lab with his feet up in front of a computer, drinking a cup of tea.

"The computer is doing the hard work," he says, pointing to his screen. There I see a graph of the water we are passing through, although the only indication of its substance is the helpful tub-shaped perimeter of the image. Otherwise a cascade of coloured numbers pours down the screen — turquoise for cold, amethyst for warm. Nils points to a sequence of purple digits, "That's a spike in thermohaline temperature. They happen when there's a freshwater current. We're seeing a lot more of these now."

I don't know if he means *now* as in an age of global warming, or now that we are passing over the Convergence. Freshwater currents in these latitudes means melting icebergs.

I look at the torrent of numbers. "I wish I knew how to read these."

Nils shrugs. "It's easy."

"For you, maybe. You've got a degree in physics."

Nils gives me a puzzled look. I don't think he's been in the company of non-scientists often. He finds my ineptitude perplexing. I am not in my sphere of power here, I have to keep reminding myself — if such a place exists.

The Antarctic Convergence is the real border to the Antarctic. Formed as the warmer water of the temperate zones collides with a cold Antarctic current, the cold water triumphs, placing a cordon around the continent. Convergence water is five degrees colder than sub-Antarctic water; it is the gatekeeper to another threshold — the realm of icebergs. Full-blown bergs are less commonly seen beyond the Convergence's perimeter; the water is too warm, although remnants of bergs have been known to venture as far north as the Falklands.

Later, Max appeared in my door. Just as he left abruptly, he never seemed to arrive; rather he materialized, like a spaceship. A dark spiral turned at the heart of his every manoeuvre. It was unusual, I thought, for someone so light and blond to have this saturnine quality.

Max, Nils, and Emilia were so much younger than me, but it didn't seem to matter on the ship, in this new Antarctic world. My relative ignorance of the details of the science they were involved in had brought us to a level playing field, perhaps. More than the others, Max embodied the benefits of youth — optimism, curiosity, energy, confidence. I want to start again. The thought arrived like Max, unannounced. I batted it away; I wasn't quite old enough to fall prey to mid-life crises.

Max's life, from what I knew of it, was already at its apogee: a parade of skiing holidays, family birthdays, music festivals, and, lately, invitations to give papers at foreign colloquia on climatology. When I was Max's age, twenty-three, two years after turning my back on my life in Canada, I had lost my parents, a job, and people

I loved. I had panic attacks on London buses, forcing me to rush off the bus and sit on benches in unfamiliar neighbourhoods — Cricklewood, Willesden — kilometres short of my destination, breathing hard for hours before consulting my *A–Z* and taking a bus home. In my spare time, I underwent brain and MRI scans for a mysterious and debilitating illness.

I told him some of my story, in part because he asked with that jackhammer insistency of his: *And then? Why did you do that? Who died next? What did you feel?* He listened with his head slightly cocked to one side. I could see his intelligence weighing up factors of fate and probability, fractional sufficiency and velocity. He was turning me into an equation: stress + bad luck = moroseness. There was a detachment to his curiosity, but also a democracy. He subjected others, male and female, to the same scrutiny, I'd noticed.

Through the hours of ship-bound longueurs we talked: in the bar, at the breakfast table, out on deck when we both exited the cosseted air of the ship to revive ourselves by being blasted by the minus-five wind chill factor. Max and I discovered we shared a vision — mine imaginary, his actual, courtesy of computer modelling — of a cold, slumbering earth, trapped in a winter that would last ten thousand years.

On the bridge, we stared out into the sea, willing icebergs to appear. We had been on alert all day and now that we had crossed the Convergence the tell-tale yellow fragments were beginning to show on the radar, just up ahead.

"What is it you find so fascinating about ice ages?" I asked.

"That the planet was completely different, then. That there were no humans, or at least not until the latter chapters of the ice age. And because it's a challenge, basically, to recreate the ice sheets. I need challenges."

I had come all this way on the lure of just such a challenge. I wondered if I should say so. Max was astute; he would not like collusion of interests, a safe strategy for establishing common ground between people, as much as collision. It would be a complex manoeuvre, to be his friend.

"I'm changeable, essentially," he had told me the night before, in one of our long discussions after breakfast, before he went on shift in the lab.

"We all are," I replied. "We're like the weather."

"I mean I can't seem to feel wholeheartedly one way or the other. Once I locate what I really feel, I'm always looking for a reason to turn away."

The freedom to turn away from the love of God. That was how St. Augustine defined perversion — its original meaning had little to do with the sexual — as an instinct in one's character to go against the grain of positivity or succour.

"I can see that. The question is why. What are you hoping to accomplish with so much contrariness?"

He looked troubled. His mouth swung downwards. "Do you want to go for dinner?"

"Sure."

"Ok. Give me five to get suited and booted."

I also got changed. Dinner on the ship was formal. It was a nightly conundrum, cobbling together a decent outfit from a wardrobe of mainly fleeces and jeans. "Oh, I wouldn't presume to tell a woman what to wear, ever," said Mike the purser, when I asked him for tips as to the dress code. I settled for black most nights. Black top, black skirt: a polar widow. It gave a sense of occasion to the evening meal. In our jackets and ties, we all looked much more competent and professional. On the other hand, a hint of the *Titanic* wafted through so much evening finery worn while icebergs lurked outside.

I waited five, then ten minutes. I wondered, did Max mean for me to wait here for him, or up in the bar? Did he mean for us to meet outside the saloon? I looked around the corner and saw that he was already seated at a table, eating. The table was full. I picked up my personal napkin with my name on the silver ring, and sat at another table with several of the ship's officers.

I was subdued at dinner. Max could have been banking on the basic principle of being on a ship — eventually you will run into each other anyway, who needs to honour assignations? Or was this a power play of some sort?

I didn't talk to Max at dinner, or later, in the bar. The dining choreography episode had depressed me and I couldn't muster the requisite jollity for the evening bar banter, so I retreated to my cabin with a bottle of wine.

I thought again of our conversation the night before, when we'd lingered at the dinner table, Max and I, long after everyone else had retired to the bar or to their cabins.

"I never look back," he'd said.

"What do you mean?"

"I never look behind me. I don't do regret."

"I don't think it's possible to run forward through life, with no reflection. How can you know where you're going if you don't know where you've been?"

His mouth was tight. "It's not that I forget. I just never think about things once they've happened."

"Thinking and remembering are one in the same." I told him about Odin's two birds, Huginn and Muninn, whose names translate as Thought and Memory. "They were ravens who flew around the world every day. On their return they told Odin all that they saw."

He scowled. "It's not possible to fly around the world in a single day. Not even an albatross can do that."

"It's an allegory. It's symbolic. It means that for the Norse, thought and memory were one."

He absorbed this. "Is that why you wanted to be writer?"

Since his comment in the Falklands I'd avoided the topic. But now I found myself opening up again. He invited confidences, despite his sternness, or perhaps because of it.

The ship hummed away underneath us. This was another detail we had become used to, so quickly, in being in constant motion. The vibration and murmur of the ship had become reassuring, amniotic, even.

"Often writers have lives that make them think too much, from an early age."

"About what?"

"About —" I couldn't think, suddenly. "Error, morality, chance, and happenstance. How much we are just biological marionettes, convinced we have a conscious life, which gives us a right to life, or so we think. But this is a delusion."

He was silent for a minute. "I haven't read many novels, but the ones I have read have been very —" He paused. "Sobering. As if they bring to life all the things I normally try to avoid thinking. Things seem to go so wrong for people in novels, as if it's required of writers, to put their characters through terrible . . . ordeals."

"Novels are about things going wrong mostly," I agreed. "If they went well the story wouldn't be very interesting."

I sat back in my chair and fell into a silence. I had written these books myself, ones that puzzled over how things had gone so badly wrong. This frequency of melancholy was so natural to me I'd hardly ever given it thought. Now, though, I found myself exasperated by my insistence on wading in the undertow of regret. Usually people did not lure me beyond the boundaries of my commitments to myself, but somehow Max was able to draw me out, on almost every topic.

"Now here's a mathematical equation," I said. "Fear attracts fear, and optimism attracts luck."

"Do you really believe that?"

"I've seen proof of it, over and over again. That's one of the reasons why it's so hard to recover from episodes of fear or despair, particularly when you're young. They take the luck energy out of you. And you might never get it back."

He gave me a stark, unreadable look. As with many looks I received from him, I still remember its charge, its complexity. For Max I might be a taboo generator. Someone who says things he has never thought before, never wanted to. I was not sure I wanted to play this role of dank Sibyl.

I have always preferred people who were smart, well-spoken but cold over people who were slightly less astute but kind. I didn't approve of this tendency in myself and was trying to rewire this instinct. You don't know him, I told myself. It's only an illusion of friendship, or even intimacy, because you have been thrown together in this way, because it has been so intense. I couldn't believe this life began only less than three weeks before. It felt illicit, like a drug addiction or an obsessive disorder. Perhaps it was our isolation, the fact that we were cut off from the world, from friends, that made what took place on the ship feel so all-consuming.

I felt more alert and alive with Max than I did with anyone else on the ship, but he also ignited a small burning unhappiness at the centre of my being. Something within us was so similar it recoiled from recognizing itself, like a cat facing its visage in a mirror.

I was aware, too, of a vertigo instinct in myself: I am prone to throwing myself into experience as you would jump into a swimming pool, just to see what happens. In Antarctica, hurling yourself into friendships carried a danger that would normally apply to love affairs. You needed to have the right friends at the end of the world. Away from your home, your family and dominion — your sphere of power, to use the phrase that occurred to me earlier that day — the smallest disharmonies could knock you off balance. And there was nowhere to run away to, nowhere to go.

I remembered the warnings of the Antarctic veterans at Conference: *People just lose it.* Their amazed delight that stolid citizens, people with PhDs and mortgages and families, could be so undone. The propensity of the Antarctic world to provoke recklessness and bad judgment had something to do with our lack of protection, I reasoned. There was no search and rescue cover in these seas: the radar screens on the bridge showed a tundra of water and ice. "The nearest ship is five hundred kilometres away, presently," Captain David informed me when I went up to the bridge later, as if on cue.

To want to experience intimacy — whether of friendship or of love — in a place so hostile to human existence is only natural, although the Antarctic is not so much hostile as aloof. Aloofness in places as much as people elicits in me a restless quiver to animate them to feel something. *Anything*. Also to exercise dominion over happenstance, to declare to the rough indifference of experience that we are alive. Love ought to be a mirror and a riposte to the Antarctic. To discover love, or even mere affinity, would honour the thrill and illogic of moving across space with an empty continent as a destination. Here, love would not be only human but also planetary; it would ignite this vortex and flourish in its barren heart.

Every day I went up to Monkey Island, the observation deck above the bridge. The deck bristled with spires and aerials for the radar and the anemometer and a giant revolving egg which turned out to be the satellite transmitter.

"It's not good to spend too much time up there," Roger the communications officer told me when he spotted me coming down the metal ladder to the deck. "The VSAT emits serious microwaves. You might get scrambled."

If the microwaves didn't kill me the cold might. We had left the sun behind just south of the Falklands. For the past three days we had sailed under overcast skies. The sea responded in kind: suet-grey, whitecaps foaming, jostling toward the horizon with a serene urgency, as if they had important business to conduct just over the lip of the world.

Waves — the word promises a tidy progression, a steady undulation of sea far into the horizon. But the sea is anything but metronomic. In reality the sea surface is a coil of opposing forces, a writhing, as if a thousand snakes were spun into each other, all determined to go in different directions. In rough weather particularly the sea ceases to be a single entity but a series of vortices and eddies, a succession of varying intents.

It is well established amongst those who study wave dynamics that the progression of ocean waves conforms to a cycle. In particular, the ninth (some say the tenth) appears to have more power than those before or after. After hours of counting on Monkey Island I was amazed to discover there really was a pattern. The waves rolled in, identical; the ship punctured them; bow spray flew up; there was a dull thud as the hull dropped down to meet the surface of the water. The deck would disappear, leaving your feet to catch up. Vertigo in stomach, roller-coaster nosedive. *Hoo!* the ship said. But then, just as you were getting the hang of gripping the bar and bracing yourself against the ship's declension, a larger wave would appear.

The ship would square itself to this aberration as you would face a reckoning. The hull knew what was coming. It levered itself higher than with the other waves. Then came a moment when a tug of war took place between the ship's steel authority and gravity. The wave pulled away behind us, leaving its empty trough. A second of suspension, then the ship would lower itself, groaning, onto the space the wave had just vacated. *Ahh — thump — crash — splash. Roar*, the ship said, as spray crashed over the fo'c'sle then drained away, rushing back to its source. It didn't sound like water at all, but as if a glass window had just shattered on deck. Then the cycle began again. This symphony was at a different pitch than the waves that came before, or the waves that came after. I took to counting these interregnums and found that they really were every ninth or tenth wave.

That evening the sea changed. It calmed, as if commanded.

"Icebergs," Captain David said. "We can't see them yet, but they're on the radar." He pointed to scattered yellow wedges on the blue radar screen. If they are big enough, icebergs buttress the waves, muting their rhythm and cyclical effect. Ice floes, pans, and pack ice can cancel the wave dynamics entirely.

David was flanked by two able seamen, or ABs, with binoculars snapped to their eyes. "Tabs!" George the AB shouted.

This was what we had been waiting for. Tabular icebergs are one of the sentries of the Antarctic. Once they are around, you know you are near the continent.

We had been seeing the odd berg since crossing the Convergence, but they were small. The officers and able seamen kept a twenty-four-hour watch for these bergs and growlers, stationed at the front of the bridge, peering at the waves through binoculars.

We saw pinnacle bergs first. Triangle-shaped, they looked like ghost ships, their spinnakers frozen into position. "These would have started out as tabs," George explained. "They're on their last legs now." It was hard to believe, how icebergs began in grandeur only to end in these doomed scraps.

The first tabular berg was sighted at dusk, which came very late now, at eleven p.m. George handed me his binoculars. "Aft, starboard," he commanded. It was too far away to see well with the naked eye. I saw a low white wedge, like a shelf soldered to the ocean, and behind it a rose sky.

We all trooped down to the laboratory at the back of the ship to get a better look. The berg was the least real thing I had ever seen in my life. I peered at it, and still some part of my mind refused to accept what it saw.

It was a shelf. It looked not unlike the White Cliffs of Dover. I could see waves lapping at its edges. But these cliffs were as white as teeth, and lit from within, as if a platinum lamp had been placed at their core.

We returned to the bridge. We opened the door and recoiled. Ice filled the windows where sea had been. Another tab stood in our way; a solid wall loomed not more than two or three kilometres from the bow. The berg towered over us, a frozen skyscraper. The ship kept a wary distance, skirting the perimeter of turquoise meltwater which hemmed the berg — freshwater, denser and colder than the sea. Radar, while competent above the surface, does not show the submerged bulk of the berg.

White was a more complex colour than I'd thought: there was the ivory of the old berg, which had been at sea for some time and lost the bleached phosphorus glare of the ice sheet. Then the transparent albumen of the soaked ice, dusky opal, a pinkish white, the metallic blue-white of an electric current or a lightning strike, a pale, dull jade. The sky was white, too, and streaked with icy clouds.

We passed the citadel of ice and suddenly the sea was covered with trapezoids and polygons, open leads snaking between them. On the edges of the floes were substances which I would soon learn were called *granite ice*, *diamond ice*, *cloud ice*. They looked like rubble, or porridge — small lumps of ice constructing themselves, metamorphosing from water into steel beams, termite mounds, petrified trees. Their shapes harassed me, demanding comparisons.

Martin the first officer told me the Antarctic bergs are often ship-shaped: broad at the bow and the stern, an optimal shape for being blown by wind. He mentioned an iceberg seen in 1927 which was reported to be 160 kilometres long. The largest iceberg in recent history was tracked in 1965; it was 140 kilometres long and had a surface of 7,000 square kilometres — roughly the size of Belgium.

No matter their size when they begin life, once adrift, icebergs' disintegration is swift. The processes of melt and erosion are so relentless that once an iceberg breaks free of the pack ice, it fragments very quickly. When it leaves its only home and sets out on its reckless quest, its days are numbered. Within two months at sea all but the largest icebergs have disappeared.

At midnight my cabin filled with itinerant shadows. I turned off the fluorescent overhead light and watched them gather. Night seemed to know its time was up. Soon — within a day or two — we would pass beyond darkness. Would we miss it? Night is so much the realm of memory, of protection and concealment. We might feel released as it melts away over the curvature of the earth or we might feel fatally exposed. I realized that night is a dimension, an occult time of regroup and repair, necessary, perhaps, in order to face the scrutiny of day. Unceasing daylight might fray our edges to shredded jute. Soon we would live in a sunlit present tense. I would not see night again for nearly three months.

I wake out of a long sleep into a world of white ashes. Black trees against the sky, black birds threading through their branches. White snowfields, last year's hay sticking out, sparse and spiky, like tiny plinths of brown glass. I am left alone on the veranda, swaddled in blankets. They had done this for generations — putting infants outside in temperatures of minus thirty to toughen them up.

At ten years old I still do this — cover myself in the rabbit fur blankets I have made from animals I have trapped and skinned myself. Our skies are empty. No planes traverse them. There is no roar of traffic, apart from the rip of the odd car that winds its way up the denim tarmac of our road. But there is a hum, barely audible, a fizz at its edges. This is the sound I will not hear again until I go to Antarctica — the sound of blood coursing through my veins, the low-frequency electricity produced by a body.

The hayfields are dun. The leaves are magenta but the earth is a defeated sallow yellow, clammed up for the year. It is October. The land awaits its annual devastation of winter.

Autumn is our time of industry; from September to November we stockpile; autumn is hunting season; every day the air is pockmarked by the retorts of rifles. At eight years old I know how to hold and fire a gun, at ten I am allowed to kill.

My great-grandmother is the best shot in the family; when my grandfather was my age she taught him to shoot. Now it is my turn. First she backs me up against a tree so that I won't be knocked over by the recoil. She teaches me to clean and load the rifle, and how to keep ammunition safe. One of the first things I learned to read was the gauge on a bullet box.

In autumn there are rabbits to be skinned, partridge and grouse to be hung and plucked. These too will go into the freezer. In the garden plums must be harvested and made into jams, pies, muffins, cakes, conserves, just as in the summer we had done the same with strawberries, raspberries, blueberries, and blackberries, in that order. All this we have learned from my great-grandmother, who is as competent a woodsman as any man, as well as a resourceful homemaker. It is she who crochets our rugs, sews the quilts, skins the animals for the furs we sleep under.

Even now that my great-grandmother is a knock-kneed old lady, I fear her. Something inside her is missing, some essential ingredient. Like a missing limb, she can survive without it. She is amusing in the way that all my family members are — characters on a stage-set, given to contradiction and confrontation, but even at ten years old I can tell she is a strategic, manipulative character who seeks purely her own advantage.

I hop off the veranda, ready for my next lesson. I follow her; she has a rifle slung under her shoulder and is dressed in a lumberjack shirt. She smokes a pipe. From behind, it is impossible to tell if she is woman or man.

The land is making strange sounds. It creaks and sighs. A hush has befallen it, not a voluntary one. Somehow I know no good can come out of the long, silent winter to come.

I am eleven. The veranda is gone. The house is gone. These things have been whisked away from us like a magician's trick, the one with the tea set and the tablecloth. We are all that is left. We are the table.

My grandfather has terrible car karma. He buys a succession of second-hand cars — Ford this, Chevrolet that — with something wrong. If it's not the steering, it's the alternator. Cue many hours spent at crossroads with the hood up. We have to hitchhike home. We make a great vagabond team — he gets maximum mileage out of the driver while I charm them.

Other times he disappears and we discover he's been up with the Buddhists. I am not sure what Buddhists are, but they have established a monastery up the cape. Apparently our province rests on very powerful ley lines. He tells us this when he comes home from his mystery weekends at the monastery, where he has a whale of a time playing the harmonica and telling stories about the days when entire families — horses and dogs too — were lost through the ice on the inland sea where we live.

The Buddhists drop him off, sober, on Monday mornings. By then my grandmother is convinced he has been dead in a ditch somewhere for at least two days. He brings home the excellent macrobiotic muffins the Buddhists bake.

Other times I find him in the bathroom, applying mascara. He had most of his eyebrows and eyelashes singed off in a tank in the war. This was in Sicily — he was the only of his buddies to get out of that tank alive. He liked to apply mascara to accentuate the sparse strands that were left.

"It looks good, don't you think?" Frankenstein turns to me. These are the days before waterproof mascara.

"I think you should use less, maybe," I say, dabbing away the rivulets of black with a Q-tip.

"You know so much about damned mascara, you put it on me."

"There," I say and hand the wand back to him.

He looks in the mirror and grins. He will do this later too, when he has lost all his teeth — take out his false teeth and grin ludicrously in the mirror, making faces at himself with his gummy, suddenly old-man's, smile.

6. THE ORIGIN AND EFFECT OF WATER

pressure ice
Ice having any readily observed roughness of the surface. Ice that has a history of disturbed growth and development.

DECEMBER 10TH

I hesitate on the threshold of the door to Max's cabin. He has summoned me by remote, sending me an email from his cabin one deck above.

He has told me more about his life in the past few days. He tells me he spent the year between undergrad and graduate school working on "luxury yachts" in the Caribbean. As an experienced sailor, he was "a crew member, not just a glorified waiter. I did some of the sailing."

"How was it?"

He shrugs. "The people were rich. The women were hot."

"No, I mean, how was it, really?"

He scowls. "I suppose it wasn't that illuminating. Or not as much as I'd hoped. I got to do a lot of really cool sailing. But otherwise it was monotonous. Very wealthy people are so boring. I missed my mum, actually."

His mother is a lawyer. She works for the Red Cross in Geneva; currently she is in Sudan, he tells me, helping create a civil code for women's rights. He is very proud of her. His father he is more cagey about. "He's in Dubai, now. I think." He tells me about his sister who is several years older. "She's doing a *stage* at the UN in Nairobi."

"What do you tell your family about the Antarctic, so far?"

"That it's beautiful. And cold."

Hot. Cold. His language is Iceland, fire and ice. Yet he is unmoved by experience, largely. What will animate him into passion? There is an inert quality to his being; he is daring experience to prod him into motion. He seems committed to finding everything underimpressive; even the corners of his mouth express this, how they turn down very neatly, like bedsheets in an expensive hotel.

This kind of brazen confidence ought to have thrown a dead zone around him, a sapping energetic field. But there is a countercurrent: a strict generosity, a willingness to talk to anyone, a sense of wonder about how the world works which is fresh and untainted by cynicism. Only under interrogation about his life or his feelings does he become inert, determined to reduce experience to unsatisfactory rubble.

The ship was sawing through the remote twilight we would come to equate with the Antarctic over the next few days, an interstitial light, neither day nor night. The screen of Max's laptop showed lines and lines of hieroglyphs; sequences of letters, numbers, and symbols known as computer code.

Again we talked about our shared vision of a cold, slumbering earth trapped in a winter that would last ten thousand years. Humans lacked the lifespan to have witnessed this complete seizure, although modern man saw the tail end of the last ice age slightly less than ten thousand years ago.

Would there have been an observable moment when the ice sheets began to retract their long fingers, raking them over the land? Would those early hunter-gatherers have been aware of how the climate balance had shifted, could they have intuited the repeal of their hunger? The caribou would return, the seas retract, exposing shoals of flopping fish. Was there an observable turning point? This was the focus of Max's calculations.

I thought of how, as a literary form, the novella flourished in nineteenth-century Germany. Literary lions of the day such as T. W. Schlegel and E. T. A. Hoffmann made their names on this slim conundrum that hovered between a short story and a novel. The literary critics of the age identified a common denominator in this new literary form, at the time called a nouvelle, the French word for "news" and also "new." The novellas rotated on a turning point, the critics noted: in German, *wendepunkt*. There was an identifiable moment in these works that you could locate and which both changed and redefined everything that came after, recasting the events before it in a bright bath of aftermath.

I realized that in my own work as a writer I was following this principle, more or less: a story or a novel should have a moment which, if you could rescind it, would recast all that then happens. A moment that rewrites the future.

"Can you show me a picture of all this?" I asked Max. I was hoping to see a three-dimensional model, not just numbers.

"Of course." Max switched screens and a glacial, watery landmass came into view. Beneath the visual model was a list of categories: *sum of separate drags, sliding velocities, shear stress*. He had written more categories: *debris in ice, subglacial water pressure*.

"What's this for?"

"The chapter I'm writing about the origin and effect of water."

The phrase stalled in my mind. *The origin and effect of water.* Max's glacial, waterlogged landscape was familiar. I saw conifers, bogs, gneiss, schist. The gouged peninsula province I had come from. Clouds of summer blackflies, moose, wolves. Then came a disorderly succession of images, like a film which has been hacked apart. They were of the trailer we lived in, after we were evicted from our grand clapboard house. Houses burning down, car accidents, funerals, abandoned houses, more car accidents, drinking, more drinking, arrests or near-arrests, miraculous reprieves, escapes in the middle of the night.

If I had had Max's knowledge, what would I have become? To be jealous of someone's knowledge is perhaps not very different and no more noble than to be jealous of their possessions. I studied the hieroglyphics of his mathematics equations again. I am too old, I thought, I will never know what they mean.

"I'm going to the salinometer room. Want to come?"

I was overtaken by a sudden gust of loneliness. I didn't feel like returning to my cabin, populated as it was by postcards of Shackleton's *Endurance* expedition and its mournful sled dogs watching as their ship is devoured by ice.

In the salinometer room, Max sat on a stool in his typical eagle posture, alert, long limbs tucked underneath him. I stood with my back to the sink, legs planted wide apart, for ballast. Our knees absorbed the shock of the hull's contact with the concrete surface of the ocean. It felt as if the steel of the ship were only the thinnest of membranes.

"Why does water freeze?" Max interrogated me.

"Because it gets cold."

He scowled. He wanted a real answer. Max then began to tell me a story which still perplexes me.

Ice crystals are of course made of water; but here, on this simple scientific terrain, we encounter a mystery — one which governs not only the behaviour of water, but of ice, and by cause and effect, everything in the planet.

Crystals begin life as water. But in an age of nuclear fission and particle physics, the way water behaves as it freezes is still a puzzle. When they freeze, most substances shrink and become denser. Water is unique: it expands and becomes less dense. If water did not behave this way there would be no life on earth; if ice were more dense than water it would sink to the bottom of the world's lakes and oceans, choking them with ice, allowing no room for the water that has given birth to and supported multiorganism life on the planet for billions of years. Because ice is less dense than water it floats, which means that water freezes only when it comes into contact with the bottom layer of ice, so freezing from the top down, and leaving the depths of oceans, even at very cold temperatures, free to sustain life.

Max and I looked at our watches in the same instant, but we hardly needed them. Our institutionalized stomachs knew the time, six twenty-five p.m. In five minutes dinner would be served. We bolted from the lab to get changed.

That evening we were again separated. Max sat opposite the captain, on whom he tried out his interrogating intelligence. I couldn't hear what they were saying, but I watched the body language. Captain David bristled slightly at Max's lack of deference. I wanted to sidle up to Max and suggest that annoying the captain on an Antarctic vessel was not a smart move — he might be put ashore on a penguin colony or to man a lone radio mast until another BAS ship returned in a year's time.

As the main course of seared scallops was served, icebergs loomed in the saloon's windows. There we were in our finery, drinking a bottle of Macon-Villages, watched over by portraits of the Queen and the Duke of Edinburgh, but if the ship were to founder on an undetected flank of iceberg we would be instantly pitched into a struggle for survival. I pictured us upended, me in my black trousers and boots, awash in a cold soup of polar water, ice, and vomit, huddled in an inflated flotation raft.

The ship wound through the alley of bergs. They slipped past, their flanks amber with the reflection of a perpetually setting sun.

DECEMBER 11TH

I wake in the middle of the night to the churn of the bow thrusters. We are stopped again. Under normal circumstances a ship stopped in the open ocean means disaster, and even a landlubber's body seems to understand this on instinct.

Every sixty kilometres or so, the officers are required to hold the ship as still as possible so that the conductivity, temperature, and density of the water column can be measured. The repeat hydrography of the Drake Passage is an annual study of the water column of the Antarctic Circumpolar Current. All the oceanographers call this journey a "transect." It's examined at the same time at exactly the same places each year.

The coordinates are taken from GPS and stops — called stations — are marked on the Master's chart with small X's.

The engines are cut. The waves don't like our stillness. They slap and bounce against the hull, the greedy note to their attentions which is absent when we are in motion. I have a sense of what it would be like were the ship to founder in these seas. Those videos I watched before I left London, nights Googling "freak wave disasters" and watching YouTube videos titled "Terrifying monster wave hits cruise ship." I remember that on average each week two ships disappear, globally. No one knows what happened to them. If they were aircraft, it would be a major story. Freak waves are becoming more common. The place they occur with most frequency: the Southern Ocean.

In the UIC Lab (whose full snappy name, the Underway Instrument Control Lab, we never used, for obvious reasons), I found Nils in front of his computer, feet propped up on a table once again. A different river of numbers cascaded from the top to the bottom of his computer screen, too fast to read.

Nils was two years away from completing a PhD on carbon dioxide uptake in the Southern Ocean. Trained as a physicist, like Max he wanted to apply his knowledge to understanding climate change, so he'd switched to oceanography.

"Why are we doing these CTDs?" I asked.

"Because the Southern Ocean plays a major role in the ocean-atmosphere climate system. Basically, if it didn't behave as it does, the Gulf Stream wouldn't exist, or many other currents for that matter. We're trying to understand if the Drake Passage is warming, long-term, and if so, at which level in the water column."

Nils explained all this in his calm, future lecturer's manner. Nils was so much more a likely friend than Max on this journey. He was a meditative, pleasant person. He was prone to borrowing the books I had brought as references, however obscure, reading them and inserting those Post-it index markers at passages that struck him. He was capable of remorse, as demonstrated in conversations about past relationships gone wrong, whereas for Max remorse might as well be a tree hyrax or a rare chameleon. Yet somehow our friendship, while cordial, never blossomed into real understanding.

Instead it was Max, incendiary and moody, who had become my unlikely confidante. He dropped into my cabin with unbidden cups of tea. He revealed his uncertainties — that he had them at all reassured me he was mortal. But confidence came at a price. I know he thought me peculiar, eccentric. Like many young people, he felt a distrust in the face of my years on the planet, as if experience was something unsavoury. He cast me suspicious looks, as if I might be his secret enemy, and so he needed to hold me close with these daily invasions of my space. If I approached him when hunched over his laptop in the lab, he stiffened and switched the screen.

"I'm not working for MI5, you know."

The shutter came down on his eyes. "I'm naturally secretive."

"As if that's something to be proud of."

It was too early to have an outright confrontation — we were all going to be on this ship for at least another ten days — so I walked away.

I wandered around the ship with my notebook and mini-disc recorder, writing down gnomic phrases from the Master's Night Orders book, from the Admiralty charts of the Southern Ocean, taking photographs of the instruments and GPS readouts, of the radar screen with its little yellow wedges scattered in a field of midnight blue.

All day and all night the CTD went down, slowly, and came up equally slowly. Max's task was to extract the canisters once the CTD came back on deck and measure the salinity. I tried to follow their conversations, Max, Emilia, Nils, but they spoke a different language, one of ghosts and stars, of shifting mysteries: *fluid equations, finite difference method or spectral method, flux correction and linearization, transient climate simulations, zonal differences, polar night jets.*

The distinction between night and day was being erased. We sat in the UIC Lab at night, one, two, three in the morning. Darkness failed to make an appearance; now night was only a murky ermine light.

I wondered, Is there a night boundary, an invisible meridian, a real place which is the threshold of night and where we run past its edge and into perpetual day? I thought of the nightmaps on planes, the ones that show darkness moving across the world in a parabola. The edge of night lapping against a continent, then creeping inexorably across it. A frontier between two countries. One might not speak the other's language, or allow re-entry once the border is crossed.

"Come on, we're goin' for a drive."

"Where?"

There is no *where*, no destination. He drove to put some distance between him and turmoil, to attempt a pre-foiled escape from the giant continent that bound us. They all did. I would do it too, one day.

We'd seen the signs earlier that night. We could tell if we might live or die by the angle he parked the car, by the message written in rubber burn marks on the highway.

There isn't a lot of room to hide in the trailer to which we've moved since being evicted from the clapboard house by his mother, who'd seen a better opportunity and sold it to a couple of artists from Boston.

I am under the bed, my grandmother barricaded in the living room. He breaks down the door. I hear the sounds of furniture crashing, her screams. He tugs me out from under the bed with one arm. I slide out on a carpet of felt-like dust.

Through the door to the living room I glimpse my grandmother. She sits in a heap on the floor, blood tricking from her mouth. Her glasses lie on the carpet, shattered. Her leg is folded underneath her body at a peculiar angle.

She's dead, I think. She's dead and now there is just him and me. He will wait until I am bigger and then I will be that heap on the floor.

It is always the same with these journeys. Still in my nightdress, he shoves me into the passenger seat of the car. No seatbelt. I am never scared. Stars, tops of trees, stars, tops of trees — these pass in a blur until they are the same substance, hewn from jewels and shadows.

He drives the darkened highway, swerving from one side to the other — not because he is so drunk. He is an excellent drunk driver. No, he's bored. On one swerve he hits the gravel, overcorrects, and we are spinning into the ditch. It's like the tilt-a-whirl I ride at the North Sydney exhibition every fall.

I am catapulted into the windshield, then thrown back onto the seat. I keep my body limp, like the rag doll I have left underneath my bed where I had been hiding.

Beside me, he is still. I begin to panic. How will I get him out? How will I get home? It is the middle of the night and I am not sure where we are. I am not yet big enough to see through the windshield without standing.

After five minutes he comes to.

"What the?" His hand goes up to his jaw. "Ha ha," he chuckles. "Not bad, eh?"

"How are we going to get home?"

"Don't you worry about that." He is still rubbing his jaw. "Don't you worry."

He says someone will come by, but no one passes. Hours later we are still sitting there and I am very cold. He takes me in his arms and squeezes.

He carries me home that night, walking for two hours along a dark highway. When we return, my grandmother is sitting in the chair. She has put her glasses on, even though the left lens is shattered. She looks at us from behind this spiderweb. I can't see the expression in her eyes.

We survive these episodes, which sprout to life from time to time like mushrooms in a neglected corner of a field. He brings home a paltry pay from the Naval yard, then the coal mine. When that is shut down he accepts a veteran's pension. He keeps busy — fishing, smoking the fish, setting traps, hunting, skinning, dismembering, freezing. Drinking.

Everyone drinks. No one drinks quite like him, though. The others are anaesthetized drunks, drooling and snoozing at the kitchen table in the morning, their feet so cold they have to be revived with tepid water, like frostbite victims.

Drink releases a demon inside him. It is not the friendly, commuting-spirit sort of demon, not a spiritual accompanier, nor even the dark side. It is a red-winged dragon.

Picture a young girl holding a rifle. It is aimed at a demon who sits at the table, his face a cloud of destruction, his fists bleeding from where he has punched a hole in the wall. The demon won, and the wall gave way.

The girl casts a wary glance at the scene. Behind the wood panelling are red wires, probably telephone wires. They dangle like shredded nerves. She will need the telephone, later.

The red dragon slobbers. The dragon has a limp left ear, a flap of skin on which she can see the veins and capillaries rushing blood to its extremities. She can see the chaos foaming in its lungs. The dragon is only good for absorbing and producing this entity, an amalgam of fury and escape. When it hits you it is like a gust of wind, but from above, as if fallen from the sky. She feels some sympathy — she will have a dragon too one day. The raw anger it will drink will smell like beetroot juice.

Violence is a cloud, she thinks. It darkens the horizon, it comes and goes on self-propelled logic. She can no more fight it than she can fight the weather.

So, a strange tableau, but not uncommon in that country at that time: a girl in her nightdress, a badly upholstered sofa and rocker set and avocado bath fixtures (it is the 1970s, after all), a .303 in her hands. Despite being roughly the same length as the rifle she is perfectly capable of shouldering it and pulling the trigger.

His head lays on the Formica table, his body in one of the stuffed chairs with their akimbo metal legs. The telephone is from the 1930s, a party line, still. The girl picks up the phone and dials her grandmother, who is waitressing at the Puffin Inn. Despite the shredded wires, the telephone works.

She talks with the muzzle of the rifle pointed toward the floor as she has been trained to do, unless you are ready to shoot. Her words echo in the avid ears of listeners-in. "You can come home now."

7. UNCHARTED WATERS

ram

An underwater ice projection from an ice wall, ice front, iceberg, or floe. Its formation is usually due to a more intensive melting and erosion of the unsubmerged part.

DECEMBER 12TH

We sail into a shifting matrix of sea ice; the ice is carved into trapezoids and hexagons. It looks still, but as we approach we see it is in motion, a gelid progression that requires the ship to haul back from time to time and nudge its way through this moving geometry.

There are more and more bergs now. When they meet they groan. It's a sort of ice language, I suppose. They sound like disaffected whales. The air stings. A new note within it, not cold, or not exactly, more like desiccated air.

No night now, only a sullen twilight. Something is alive but hidden inside it, a lazy grey-opal animal. Grey, blue, chalk, silver. Why haven't I seen this colour before? Something in me responds to this dimming, a peculiar urgency. Desire to live, at any cost.

As we passed further into this sombre twilight we all took to spending more time out on deck, swaddled against the cold, staring with a cowed fascination at the new landscape we were entering. Nils and I shared a position on the starboard deck, a low angle from which we could observe the striations in icebergs as we passed. We saw veins of dark grit followed by planes of turquoise transparent ice so pristine and captivating we forgot to take pictures.

"It's so beautiful," Nils said, more than once. In his voice I heard that note I would hear so often in Antarctica, of hollow rapture. There was dread too, at encountering such a beautiful place but one so unmoved by our gaze. In the coming months on base, Tom would tell me, "The Antarctic is the perfect unrequited lover. It takes everything you have to give and offers nothing in return except itself, but somehow that's enough."

The association of coldness with deadness has as long a history as its antithesis, the idea that heat is vital and sunny. The psychological character of coolness versus heat is similarly well established, a metaphorical antithesis worked out very often in fiction; witness the coldness of Karenin and the duel of frost and fervour in Vronsky in Tolstoy's great novel *Anna Karenina*, or Dickens' Miss Havisham, turned into a living ice sculpture by her refusal to accept the passage of time. If heat and cold have been converted into established psychological tropes, the directly spiritual dimensions of ice have largely vanished from the cultural map of the twentieth century.

Ernst Haeckel was a German biologist, naturalist, philosopher, physician, professor, and artist. In 1895 he posited the concept of "soul-snow" — a sort of crystallized soul. He believed the soul to be bodily substance, a form of gas that could be liquefied under high pressure. If it could be turned to liquid then it could be frozen, and become evanescent, like the scattered ashes of the cremated, to flow in the wind. This was called religious thermodynamics, a short-lived philosophical vogue that sought to interpret the godhead within the laws of temperature: a cosmology of heat and cold.

I decided to find Max — he was German-speaking after all, he might have heard of Haeckel — to see what he thought of this bizarre soul-snow theory. I passed by his cabin on my way back from the upper observation deck. His door was open but his curtain was drawn. I found him sitting at his desk with his forehead cupped in the palm of his hand. His hair streamed through his fingers. He stared blankly at the desk.

"What's wrong?"

"Oh, nothing." He was not very convincing. He seemed to realize this. "I've just found out something that . . ." He paused. "About someone . . ." For the first time since I came into the cabin, he looked at me directly. His eyes were clouded with uncertainty. "I might as well just tell you."

He had been hoping to have a relationship with Emilia, the Italian student, but she had told him the night before she was to get back together with her ex-boyfriend.

"I have no idea what she is thinking. We have no real rapport. She's so —" He frowned. "Cagey."

It was there at once, the illogic of love. He knew there was no real connection between them, and still it didn't change the way he felt.

He retreated to the far side of his bunk, pulled his long legs underneath him, rested his chin on his knees. "I'm used to getting what I want. Whenever I've wanted someone, they've always been available to me."

"Well it won't always be like that."

He gave me a look. He had heard the tart ring in my voice.

"I wasn't looking for it," he said. His expression was familiar — the peculiar guilt which arises from wanting a miracle. "Love."

I sensed a trap. Not a conventional one, rather the kind of snare sprung by fate, to test your commitment to life. Anything I said now would be a cliché, which I dutifully delivered. "That's usually when it comes."

I made to leave him to his misery.

"Wait. Don't go." His green eyes had darkened to charcoal,

flecked with small bits of pain that glimmered. "Can I show you some pictures?"

He levered open his laptop and clicked through to a series of images. A dog like a lean question mark with a narrow, melancholy face appeared. "He's a borzoi," Max said.

"What's his name?"

"Bismark. But we call him Biz."

Then a swimming pool, a cool tourmaline rectangle in the midst of a lawn so perfect it looks stencilled in. A house with two turrets. Beyond one of them, the cold lip of a mountain.

"Is this your house?"

He nodded. "It's just outside Zurich. It's strange. I love being here — in the middle of nowhere — but I've never missed home so much."

A tall blonde woman appeared. Her eyes were the same colour as the swimming pool. A gold watch reposed on her wrist. She had Max's mouth, or he hers, but their faces were not the same — she lacked his severity and drama. Beside her was a photocopy of herself, the same blue eyes, the same flaxen hair.

"My mother and sister," he said.

"I guessed. Where is your father?"

"Away. He spends a lot of time in Stuttgart these days. That's where the venture in the capital is, apparently."

"Stuttgart? I thought all the finance was in Zurich."

He shrugged. "The truth is I find making money so boring."

There was a bite to his words — *so boring* — usually easily uttered, with a lime-coloured sigh. No, he was angry at money, for having taken his father away.

I glanced out the porthole. The sea had turned suet. Globules of ice accumulated on the ship's railings. The sombre cloud Max emitted had an unfamiliar density.

I realized I had been wrong about him: he felt too deeply, rather than not enough. He was not reducing experience to rubble as much as waiting for it to live up to an unuttered promise he felt it was reneging on, to enliven him. He hadn't experienced enough deprivation to calibrate his emotions properly — he had grown thick on the fat of prosperity. But his essential nature was lean, acute, like an isosceles triangle. His spectacular upbringing had put him to sleep. He was waiting for something to jolt him into life.

"Thank you," he said, "for the talk." Something had snapped shut in his voice. He was not really thanking me. He wanted to be the person to dismiss me, rather than me abandon him to his sorrow.

"You're welcome." The note of formality in my voice was new, or I hadn't used it since we first met. It rang, tinny, in the air between us.

I went out onto the aft deck, which was low, close to the water. I wore no coat. It was cold, but not unbearably so. I hung my arms over the gunwale and felt the spray vaulted up by the waves. We were moving through moderate seas for the Antarctic, light swells, waves no more than a metre high. The ship juddered as we sank into a trough. The water was black and slick; it shone like dark stone.

I imagined myself in the water, the ship pulling away, kneading the waves until it was a white beacon against the horizon. It would be a long time, perhaps the next day, before anyone noticed I was missing. I had already skipped meals in an effort not to balloon in weight (a real risk on a ship where you could eat a full English breakfast every morning and a steak each night) and no one had come to check I hadn't tumbled into the southern Atlantic.

This is how you die in cold water: first the cold shock, a heart spasm provoked by the initial jolt of immersion into zero or sub-zero waters. The spasm makes the lungs gulp. Without the buoy of a lifejacket and the protection of a mouth guard I would have inhaled water and filled my lungs within a minute. Cold shock causes the heart rate to accelerate so rapidly that you breathe desperately, trying to get enough oxygen to feed your sprinting heart. This is when most people drown, long before hypothermia sets in.

The water was so black. It looked almost solid, like licorice.

"Hey!"

I turned to see someone walking toward me: orange boiler suit, rigger jacket, woolly hat. I scrutinized his face. I hadn't met this seaman before. Where had he been keeping himself? There was a wary deliberateness in his expression.

"You're not supposed to be out here."

"I'm sorry," I said. "I didn't know." This was a lie. I knew full well I shouldn't be on deck without rigger boots, a hard hat, and a jacket.

"You'd better go inside." He accompanied me to one of the outside portal doors, as if he were certain I had other intentions.

Back in my cabin, I stood in front of the mirror. I saw a small woman, not entirely familiar to me. She was thirty-seven years old. She was shocked by how easy death would be in the Antarctic, how available.

That night I took photographs of this woman in the cabin mirror. I still have them on my digital camera. For some reason I turned on the black and white function. I moved while taking them, or the ship was rocking, so the photos show indistinct outlines: black, white, a mirror, then grey. A blurred figure, a ghost.

DECEMBER 13TH

Today I begin a series interviews with people on the ship. First Captain David, then Elliott, an Antarctic old-timer and the base commander at Port Lockroy, where we will arrive in two days and where he will leave the ship. I ask Elliott about the old days when there were dogs, not skidoos, and no women.

"I don't approve of it, really, women here." Elliott gives me an expectant look. When I refuse to be scandalized, I can tell he is put out. "Because it upsets the balance, the equilibrium we had. Men start competing for women, and then it's not so easy to be a community. In my winters we were as one —" he raises his hand, his fingers intertwined. "We were unbreakable."

"From what I've heard the mixed communities manage a similar spirit," I say. "I've had several people tell me they think women in the Antarctic are a good thing, that they bring a civilizing influence, that the communities are more emotionally stable."

"We had it good in the gold days: dogs and us. That's all we had to rely upon. There was a peace in this, a kind of glory." He gives me a valedictory look which says *you had to be there.*

In the end Elliott doesn't play the role of bigoted scandalizer well. Gallant and well-spoken, he is a mine of information about the Antarctic, and the Arctic too — as it turns out he is a champion dog-musher and lived in Alaska for years.

"Why the polar regions? What is it about them?"

"I just loved the life, the outdoors life." As he speaks the russet outdoorsman shine in his skin deepens. "I want to be outside, always — in the snow, on the sea, in the air. And we had such an important job to do, with the science, and maintaining Britain's historic claim on the continent."

I don't agree with everything he says — certainly not its imperialistic tinge — but I am taken by Elliott's wonder and conviction. How to communicate this, without sounding mawkish, besotted, especially to an audience whose emotions have been sanded at the edges by complacency?

This is a question I ask myself over and over here: Is it possible to convey these emotions in fiction, and what is the point, anyway? Sometimes life just has to be lived, not recorded, examined, re-framed in aesthetics and imaginary worlds. Is this why I feel thrilled and awake, as if I have been sleepwalking all my life, or at least for an interim — because I am surrounded by people for whom life as it is lived is simply enough?

You have made a career out of your sadness. We were in the hallway outside his cabin, waiting for Nils to come out and go to dinner together. Max came out with this rejoinder to something I had said. I recoiled. Then Nils appeared, and my chance to give a riposte vanished.

I ate at the same table as Max that night, but two or three people away. He lobbed questions at his interlocutors much the way he did with me. Perhaps our rapport was not what it seemed, rather just an ordinary engagement for him. The crucial factor was whether the other person responded; many did not. His inquisitiveness and open manner invited confidences, but I already knew these were misplaced. Why did I persist? Was I so desperate for someone to talk to?

As we were eating, David's voice, smooth as molasses, came over the PA. "Ladies and gentlemen, we are now entering the Gerlache Strait."

The Gerlache Strait was named after Adrien de Gerlache, the Belgian explorer who mounted the *Belgica* expedition to the Antarctic between 1897 and 1899. Named for the ship which conveyed them, it was the first expedition to deliberately overwinter in the Antarctic; de Gerlache's strategy was to allow the vessel to become iced in at the beginning of the austral winter, then to hopefully be freed in the spring melt some seven months later.

Aboard was a man who would make his mark on the history of polar exploration: Roald Amundsen, sailing as first mate, who would later be the first man to sail the Northwest Passage in the Arctic as well as the first man to reach the South Pole. It was also the first entirely scientific expedition to the continent.

The ship was iced in early March. On May 19, the sun set, and would not rise for another sixty-three days. From the beginning, the expedition was beset by discord, drunkenness, and ill-discipline. Although he managed to eject the troublemakers before the ship reached the Antarctic, de Gerlache found his men hounded by illness, mental and physical. One member of the expedition died of a heart attack and at least two others succumbed to severe depression. Amundsen kept a diary of the winter spent trapped in the ice, as did the ship's doctor, a young American named Frederick Cook, who would also leave his impress on the history of polar exploration when he would claim, controversially, to be the first man to set foot on the North Pole in 1908.

Cook documented the winter and the men's ordeal in an unsparing light, blaming the confines, the peculiarities of the ship's captain, and the constant darkness for the men's psychological disintegration. Cook's account of the thirteen months spent aboard the ship read in sharp contrast to the narratives of collective endeavour and relative harmony that Scott and Shackleton would later pen: "The truth is, that we are at this moment as tired of each other's company as we are of the cold monotony of the black night and the unpalatable sameness of our food. Now and then we experience affectionate moody spells and then we try to inspire each other with a sort of superficial effervescence of good cheer, but such moods are short lived. Physically, mentally, and perhaps morally, then, we are depressed."

It is a remarkably honest account, especially as the word *depressed* did not enter ordinary lexicon until some fifty years after Cook's diary. In the end, only one of the *Belgica* expedition's members died, Lietuenant Danco, a seaman whose name graced the coast to our port side as we steamed south that evening. But there was another important casualty — Nansen the cat, a great source of amusement and comfort to the men. Around June, the time when the men began to withdraw into fits of distemper, the cat mimicked them. Previously friendly and affectionate, he hissed and spat whenever one of the men tried to feed him his daily ration of penguin. Then he simply died.

The darkness was of a different order, as was the cold, than anything he had ever experienced, Cook wrote. It was as if the human body knew it and shrank away. The men's faces became green, then jaundiced. Their vital organs began to pack up. The body abandoned them, perceiving itself to be in an alien place.

That evening I parked myself behind the bar and played barmaid while Steve told me stories of living five metres below the surface in the old Base Z.

Steve was the real Antarctic item: a burly man in a checked shirt with a beard. "The base was covered with more and more snow every year," he said. "Until we had to build a staircase down through the ice and snow to get there. Our breath frosted and in the mornings we'd wake covered in a film of ice."

After Steve left, one of the ship's deck officers came over to the bar. "So now you've seen a PermaFID. That's what we call guys who come back to the Antarctic, year after year. Some of them do multiple winters, one after the other. They find they can't leave the life."

"Why?"

"They get completely institutionalized. The organization feeds you, clothes you, it insures you, transports you, it gives you a social life. A kind of family, even. There's not many places you get all that now. Except maybe the Army."

Later that evening, on his way to the UIC Lab, Max appeared in the doorway of my cabin.

"What's that term, when the weather mirrors your feelings?"

"Pathetic fallacy."

He was still trying to charm Emilia. He was annoyed with himself; this was not something he would normally do. He was acting compulsively, irrationally. He was no longer capable of making a choice where Emilia was concerned. He was becoming aware that much of what we do in life has nothing to do with choice at all, rather with interior switches which are thrown or not thrown, and the implacable intentions of fate.

I saw a former self in his abjection. Did I not chase people at his age, thinking I could apply my will to people the way I applied it to my Political Science Advanced Topics seminar? It didn't help that we were alone there, in those rough seas, the thickening ice, thrown back on ourselves with only our characters for armour. Over the last few weeks, the tether that held us to our normal lives had stretched, then thinned, then finally snapped.

So much of what we think of as ourselves is actually borrowed from external circumstance — our sphere of professional power, the liens of family, status, money. In the Antarctic, scientists are secure in their power; in fact the continent serves as a giant outdoor laboratory for science. Along with the support personnel, scientists are imported into this world entirely so they can enact their expertise, divining for frozen subglacial lakes, calculating the shear of a moving glacier, mapping rogue ice shelves which dissolve into the sea each year.

But a writer is always without a sphere of power and the social legitimacy it brings because it is only consecrated in her books, in this book, which does not yet exist and may take several years to materialize. Or it may never appear. Even Suzanne had more of a three-dimensional presence in the Antarctic world; she moved around the ship making short films, studies which would inform her future giant sculptures of ships' hulls and CTD canisters.

I had only the power of my persona on the ship, and it was not enough. I am an introvert, a thinker, a watcher — not enough of a personality, a character, to triumph without the other spokes that keep me centrifugal in my normal life. But somehow, despite his greater charisma, Max and I shared a category-less aloneness there, on a ship at the bottom of the world.

We were learning, all of us, that on a ship it is impossible to disperse emotion. There are no parks to walk in, no friends to drop in on, no shops or art galleries to distract you. I had lived without these everyday comforts before — I was raised without them — but lately I had come to take them for granted, I realized. Confinement of any kind is a test. We have to find our own ways of dealing with the tension, boredom, and irritation that come with it, especially when these are underwritten by a barely acknowledged note of fear.

As our ship climbed and plummeted the crests of waves, I thought, I can't take these highs and lows, it's too volatile for me. I felt myself to be the victim of a dark prank of an indeterminate nature but which made me feel my aloneness and the consequences of choices I had made far more on that ship than I did in London. I felt an appalled fascination. I carried this around the ship with me like a scientific specimen.

In the meantime, Christmas was coming and we planned on decorating, which, the ship's officers informed us, was traditionally a job for the FIDS, and particularly the women. But among the FIDS a lassitude had crept in. I went up to the bar at four p.m. to make a cup of tea and found a group drinking and playing Trivial Pursuit. The same group was there after dinner, at eleven p.m., at one a.m., at five a.m., by then in a state of slumped disarray. No one was rowdy or aggressive, rather crumpled and impassive. Our futures lived on the continent, still several uncertain days away, and for the moment there was nothing to do but drink.

DECEMBER 14TH

Elliott and I stand in our parkas on a rocky outcrop, surrounded by over a hundred Adélie penguins. Penguins — or rather penguin guano — smells like fish and rotten hay, with a fringe of chalk thrown in to dampen its worst effects.

"It's the ammonia." Elliott sniffs. "But after a few days we won't smell it anymore."

The guano is a pale green, like the mucus of a thick cold. It is smeared across the rocks, making them more slippery than if they were covered with pulverized banana skins. One or two people have already tripped and bashed their knees on the smooth hard stone.

"It's a relief to smell something for a change." I don't say, I can't believe you're going to stay here for four months. That wouldn't be helpful, and besides Elliott looks like someone who knows what he is in for.

Elliott surveys the exposed and isolated point which will be his home for the next four months, along with his two companions, Mike, a mountaineer still on a high from his recent successful ascent of Everest, and Liz the avid birdwatcher from Norwich, who will assume the title of world's most remote postmistress.

The hut sits at the foot of a towering glacier. Approaching base from the ship, we saw it first as a speck, a tiny beacon painted the colour of cooking chocolate, with blazing red window frames and a tattered Union Jack fluttering on a toothpick flagpole. As we got closer we could see a multitude of small black forms, slick and glistening, like leeches or polyps, lumped all around the hut. These were the penguins, who we spied on through binoculars and who wiggled their behinds and sprayed green goo in response to our arrival.

Lockroy is a lonely place, dwarfed by regal mountains, surrounded by white loaves of hummocked glaciers that spill from them into the sea. The hut itself is like a piece of Lego. My eye is still accustoming itself to the gigantisms of the Antarctic. "I can't get a fix on the size of things here," I tell Nils when we take a break from hefting boxes.

"It's a problem of scale, fundamentally," he says in his precise way. "Everything is too big, and too empty of anything human. It's just the same as how humans can't get a fix on the atom, visually — it's not an issue of belief but of scale."

During the Second World War, Lockroy was an important communications post, ferreting out the German submarines which slunk across the ice shelf edge. Now it is a live museum. Each austral summer thousands of cruise ship passengers are ferried into Lockroy to visit and buy memorabilia and stamps for postcards that will eventually make their way to their destinations, emblazoned with philatelic penguins and Edwardian heroes.

"They come just after their champagne lunches or bouts in the jacuzzi. The women wear heels! And what do they find?" Elliott shakes his head. "Three nutters in a hut in the middle of nowhere."

We had only a few hours to unload four months' worth of stamps and provisions. Postage stamps turned out to be twice the weight of tinned mushrooms. We struggled on the slippery rocks in our bulky cold weather gear.

Pitching in is the first rule of the Antarctic, unwritten but sternly conveyed in a myriad of ways; almost nothing else about one's behaviour or demeanour matters. You "muck in," and with energy, otherwise you will be termed "slack" and risk being shunned — one of the many ways in which collective endeavour is privileged over personal preference in Antarctic society. A month before I might have been working at my day job as an editor sitting at a desk in London fielding manuscripts, but suddenly I'm wearing rigger boots and being winched aboard a cargo tender among seamen, ready to offload four months' worth of provisions to an Antarctic base.

Once ashore we started emptying the tender, wending through the obstacle course of penguins, who took no notice of us, preferring to loll, squawking, on their stomachs on the pathways up the rocks. They looked like ice curling stones. We must have looked equally bizarre to them as we performed a waddling choreography of weaving and ducking, slipping on the guano, sliding dangerously close, and no one wanted to squish a penguin underfoot.

"We're ready for our summer. Over," I heard Elliott say on the radio to Base R. Elliott would report to the Base R communications manager every day, once in the morning and once in the evening, on the "radio scheds." He grinned — that ecstatic, matchstick smile I would see so many times in the Antarctic, which burned so bright for a moment, before fizzing into a black, dazed awe.

Just as we were readying to head back to the ship the weather closed in. It started to snow, stinging flakes that seared our faces. A wind came barrelling down from the glacier and the sky darkened.

This was the first time I experienced a katabatic wind. The term is derived from the Greek word for *descent*. Katabatic winds are the only winds on the earth driven by gravity and not by the rotation of the earth. Born high on the ice sheet and pulled down toward sea level by gravity, as they accelerate down the ice cap and glaciers they gain velocity. One moment it's windless and the next you're knocked off your feet. These winds regularly flip Antarctic aircraft over like toys.

Captain David came on the radio. On the ship they'd had to take the tender out of the water and put it back on deck. It was rough for it to get through to us. "We might send the RIBs in," the first officer said on the radio, referring to the Rigid-Hulled Inflatable Boats BAS used for quick transfers. "That's what they say but it's too rough for RIBs," Elliott informed us.

We all huddled in the kitchen. We knew the ship's officers were thinking about their schedule — we were expected the following day at Vernadksy, the Ukrainian scientific research station down the road, so to speak. We were already behind schedule and Captain David was thinking, no doubt, about the inconvenience of having twenty people stranded at Lockroy, waiting for a change in the weather which may or may not come any time soon.

We went to the window. We couldn't see the ship. We peered through its snow-encrusted panes, watching as the slabs of glacier were eaten by mist.

"Well, I guess I'd better start cooking dinner for twenty-five," Elliott said.

Elliott began to tell us a story. Two years before, a French yachtsman had pitched up in Lockroy. He had sailed his yacht across the most treacherous seas in the world. He ignored Elliott's warnings and went for a walk only a hundred metres from the hut. A moment's inattention meant he failed to see the tell-tale mint blue seam in the snow. He fell into a crevasse.

"He actually died from head trauma," Elliott informed us. "We winched him out within minutes. But he wasn't wearing a helmet and he banged his head. Spring ice is hard."

The VHF radio crackled to life. We heard the first officer say, "We're sending in the RIBs."

We said goodbye to the hut nutters and the penguins. I took one last look at the hut and the glaciers that surround it. They looked benign, like a pristine slope at a ski resort, with their fresh icing of snow. We were in a different realm now, I realized. The dangers of the sea would soon be swapped for those of land. Elliott's story was a warning. Life was tenuous in the Antarctic, he was telling us, this joking, boisterous horde of polar ingénues.

A second's unawareness could kill us.

That night the three Lockroy staff would cook dinner on a couple of Primus stoves, huddled in the kitchen with their Tilley lamps as the only source of heat. They would bed down fully clothed on hard bunks with cold duvets. On the ship, meanwhile, we would have dinner in our finery, we would drink scotch in the bar as the ship's doctor, Dan, administered his weekly pub quiz — being the quizmaster is part of the Antarctic doctor's job description, a stipulation inherited from the days of Shackleton and Scott and their multitalented expedition doctors.

In the RIBs we gripped the hard sides of the inflatable as we zipped through the waves. We wore no immersion suits and the third mate was careful not to hit a wave straight on. I was wet, although not soaked. I didn't feel the cold.

Ahead of us the ship rose and expanded; as we curved underneath its bow its red hull and its superstructure soared overhead, the height of a six-storey building. My heart surged — that was how it felt, a wave, a rush, from inside the muscle itself, a heady brew of thrill and gratitude. I thought of Shackleton's men, how they must have felt the same passionate admiration for the *Endurance*, which had conveyed them so faithfully to the bottom of the world, then to watch it be crushed. It would be like having

your own heart smashed in a vice of ice.

The winch came down, hooked onto the RIB, and we were levered into the air, whole, and swung onto the aft deck. The ship felt more stable than a planet. It felt like home.

That night we glided through a channel between Anvers Island and the peninsula.

"Why aren't we in open water?" I asked Martin, who had finished his watch on the bridge and come down to the bar to have a drink.

"Dunno. The Old Man must have an idea." The officers called David alternately "Captain" or, when he was not present, "The Old Man" — possibly a term of endearment, possibly a rank-and-file custom. David didn't look particularly old; he had begun his career when a young man, and that year was his twenty-seventh season in ice. It was also his last. After this cruise he would retire, and this gave our voyage a sense of finality, as if his retirement was not only a milestone for him but for the organization itself. When we reached base — at that point there was no question of *if* we reached base — there would be a party to bid him goodbye.

Caroline the diver and I went up onto the bridge and found David there, alone apart from an able seaman keeping lookout to starboard. David wore jeans, a cable-knit sweater, and — incongruously, on such a manly Englishman — clogs; with his jeans rolled up on one leg, exposing calves more sinew than flesh, he strode from chart table to console, whipping out the flag-sized pieces of paper on which detailed nautical charts were drawn. He greeted us but seemed preoccupied. I went to the chart board. Our course was drawn, as usual, in pencil — lines crossed with x's showed the distance we had travelled, and the time of our position. A message flashed on the electronic maps displayed on the ship's computers: *Warning — uncharted waters*.

Caroline and I decided to recite "The Rime of the Ancient Mariner." To our mutual surprise we could remember entire chunks of it. We parked ourselves at front of the bridge and chanted:

> *And now the STORM-BLAST came, and he*
> *Was tyrannous and strong:*

He struck with his o'ertaking wings,
And chased us south along.

With sloping masts and dipping prow,
As who pursued with yell and blow
Still treads the shadow of his foe,
And forward bends his head,
The ship drove fast, loud roared the blast,
The southward aye we fled.

And now there came both mist and snow,
And it grew wondrous cold:
And ice, mast-high, came floating by,
As green as emerald.

David interrupted us. "Okay, boys and girls, we are doing sixteen knots in uncharted waters. We need some quiet on the bridge."

Caroline and I were horrifed to be thrown off the bridge by the Captain, never mind the *boys and girls*. We left, chastened. Downstairs in the bar, we were so quiet people asked us what has happened. "Don't you know?" Mike the purser told us, "It's bad luck to recite that poem on a ship. Especially in the Antarctic."

There it was again: luck. That the Antarctic was as superstitious a realm as the theatre was not surprising, maybe. I remembered the jester duo, death and luck, which patrolled Antarctic literature. I retreated to my cabin. My chaperone had returned. For a while, while we were stationary at Lockroy, he had disappeared. Now I saw him outside my cabin porthole, a sooty albatross. He flew stalwart with the ship all night.

DECEMBER 15TH

Six a.m. There is a knock on my door. I wrap a sheet around me. My muscles ache from carrying boxes into Lockroy.

It's Max, fully dressed: parka, hat, woollies. "We're going through the Lemaire Channel. I thought you'd not want to miss it."

I get dressed: long underwear; thermal T-shirt; fleece no. 1, no. 2, no. 3; woollen socks; boiler suit; parka; waterproof; two hats; rigger boots. In this costume I waddle up to the Monkey Island.

We are moving very slowly through thick fog. I can't hear the engines. The ship seems to be gliding. On either side diamond-shaped tors of rock, dusted with snow, congeal out of the fog. They are five hundred metres high and close enough to touch.

The channel must be more of a trench, I think, if our ship with its deep keel can get through here. It is alarming to be only inches away from vertiginous scraping rock. In front of the bow is a cul-de-sac. Our exit is blocked by a glacier; its white tongue lolls into the tar-black water. We are sailing straight into it. At the last moment possible, the ship veers to starboard and the tors part to reveal a narrow gap. Beyond it we see open water and ice-coated islets.

Max reads the ice for me. "Frazil ice," he says, pointing to the splintered rime on the surface. Then, "Old ice. You can tell because of the discolouration." I see its giant flakes, like rock shale. Some are light blue, others are grey or chocolate brown with grit. "From the basalt," he says. His voice is narrow and impatient, so I leave him hanging over the top rail of Monkey Island, his eye fixed on the dark water.

The ice pans are forming. Once out of the strait we proceed slowly, at only two or three knots, as if the captain is afraid to startle the ice. It parts willingly, shifting its gruel-like soup. The sea ice glares. White is an absence of colour. Yet this deprivation glitters, it is dazzling.

White is an elusive colour. In my cabin, I looked up synonyms in a thesaurus. Many words for white are suspicious and pejorative. In humans the colour is linked with illness: *whey-faced, sickly, ghostly, poorly, sallow*. Many of its synonyms rely on naturally occurring substances or chemicals: ivory, wax, cream, zinc, titanium, quicklime. *Quicklime* — calcium oxide, a chemical compound that is caustic and alkaline. It is good at cutting the scent of decaying flesh and this was its use, for hundreds of years. Quicklime describes the snow at sea level, where the temperatures are milder and the snow granular. But on the glaciers, I noticed at Lockroy, the snow was dry and powdered, as light as coconut. In the wind it rattled, a million minute wind chimes.

White is the colour the eye sees when light contains all the wavelengths on the visible spectrum. If you pass white light through a prism it is sliced into a rainbow: orange, green, blue, red. White is all colours combined until it looks like none.

I strove to see anything in the ice field ahead of us. Its glare was painful to look at: it had glamour, but was also phantasmal. *Glace* — the French word for ice sounds more faithful to the true nature of the substance, with its brittle slipperiness. Spread out in front of us, as far as we could see, it looked less like ice than crushed metal.

That afternoon we left the ship to go into Vernadsky, the Ukrainian base. The water darkened to a thick celluloid. The mountains, by contrast, shone with a floodlit phosphorous white. We had entered a silverprint world; what was normally light was dark, although throbbing with a substanceless, unstable white.

We went into Vernadksy in the RIBs, wearing our immersion suits this time. Once again we zipped through clouds of stinging flurries. In our immersion suits and lifejackets we waddled stiffly in single file up to the base through a corridor of snow to meet the welcoming committee.

The Ukrainian chief biologist wore those glasses which adjust with light. Although the day was overcast his shades were dark. He looked like a racecar driver temporarily marooned in the Antarctic. He was dark-skinned and (I imagined) dark-eyed and wore only a fleece and stood outside in minus five with no gloves and no hat, as if he'd met us on a slightly overcast day in England.

"Well, he's acclimatized," said Peter, an oceanographer.

We stood, a group of orange overstuffed penguins in front of stern tin sheds, surveying the base. One of the oldest bases in the Antarctic, Vernadsky was first built on the Argentine Islands as a British base, during the British Graham Land Expedition (1934–1937). Perched on a rocky tip of Galindez Island, the base was once called Faraday, and in the 1980s the British, unable to either maintain or dismantle the base, as the Antarctic Treaty requires signatory nations to do, sold it for £1 to the Ukrainians, who began research there in 1995, naming it after the Soviet geochemist Vladimir Vernadsky, who also founded the Ukrainian Academy of National Sciences.

We were shepherded from lab to lab. I wasn't sure the Ukrainians had got a good deal. Peter elbowed me discreetly when the chief scientist's back was turned. "Look at that microscope!" I spotted my high school chemistry class microscope on the bench. In the labs we saw several of the original British notices stating emergency procedures and VHF radio frequencies still tacked on the wall.

The base had a meteorology room staffed with creaking PCs, a serviceable though dusty VHF radio, and a makeshift gym, one room with rusting antiquarian weights and a token rowing machine. Of all the rooms on base, only the gym struck me as particularly Ukrainian, thanks to the posters of big-boned blonde women in 1980s workout leotards on the wall.

We were ushered upstairs to the bar for shots of vodka — it was ten in the morning — and slices of smoked salmon. Behind the bar we found the diesel mechanic, dressed in a pristine white shirt and grey trousers. He was greying, with a bushy moustache and mournful blue eyes.

"How many of you are here?" I asked.

"We are thirteen. Lucky number!"

"Are you all men?"

His face slipped. "Yes, there are no women." That explained the peek-a-boo chorus of eight or so staffers who kept looking at us (we were two women amongst twenty men) from round the corners of the base corridors, as if they'd just spied an Easter egg.

Vernadsky enjoyed the reputation of having the best-stocked bar in the Antarctic, and, for being 67.15° S, it wasn't bad: Johnnie Walker, Talisker, three types of vodka, four types of gin — another triumph of the British legacy. A massive brassiere was slung between the bottles of vodka and whisky.

Suddenly a group of people burst into the bar. A woman wearing a fur coat and hat plunked a bottle wrapped in Christmas paper on the bar and embraced the diesel mechanic over the counter. Our party turned as one, astonished, as more and more people poured through the door, laughing, singing, a parka-ed, jolly herd. We had peeled off our padded boiler suits, but we still looked a motley crew in comparison to those fur-coated people, with our grease-stained jackets and moleskin trousers.

"It's a Russian cruise ship," Martin, the first mate, who came ashore as our RIB driver, answered my questioning look.

"I don't remember seeing a ship."

"They're anchored around the other side of the island, I saw it on the radar. The cruise ships like to hide, to give their passengers the impression of pristine wilderness. I've heard them on the radio arranging to sneak past each other so that the passengers don't realize there's another thousand people hot on their heels."

The Russians installed themselves in the corner and started on the vodka.

The diesel mechanic took me by the arm. "I need assistant." He manoeuvred me to the pool table in the middle of the bar.

"No, no," I protested. "I'm a terrible pool player."

"No. *Assistant.* You stay here!"

He returned with a handful of coins, which he placed on the pool table, and two coils of rope. He said something in Russian and the cruise ship passengers gathered round. "Magic show!" then more Russian. The crowd clapped and roared.

The diesel mechanic put me through my paces with the coins and the ropes. I was a poor magician's assistant, not to mention unglamorous, in my oversized moleskins and lumpy fleece. I got the rope tangled, I failed to tie the magic slipknot, or untie it. The Russians didn't seem to care, they laughed and clapped. Everyone downed more vodka.

I continue to walk home from Shades of Light, but now I avoid the railyard, where oxidizing freight cars could easily harbour an attacker. Instead I stick to side streets on which giant elms have only just come into leaf. I walk without my Walkman, passing in and out of clouds of early blackflies.

A car appears on the street beside me and slows. I glance to the side. Inside it, half-hidden in darkness, is a man. A shaft of evening sun falls across the windshield, obscuring the interior with its reflection.

"Excuse me."

I squint. The face has my eyes.

The voice comes again. "Excuse me."

I keep walking, faster and faster.

I take the exit to the Trans-Canada highway, driving past the reservoir, the lake, the dam, up the long sloping hills rising from the river valley into a sharper, colder plateau. The only other traffic is juggernaut tractor trailers which pass me in an orange blaze of light. This isn't safe, I think. What if I broke down? Those roadside emergency payphones are a long, long way in between. Walking at night, kicking gravel, I'd be perfect bait for the murderer.

A man — everyone assumes it is a man — has murdered two women in the past six months. Both women were walking alone; one on the side of the highway I drive now, the Trans-Canada, one on an isolated riverside path. Both were heading home, not particularly late, four thirty and seven o'clock.

These murders have been reported hesitatingly by the town's media. Both women were working class, one a chambermaid in a highwayside hotel not far from her house, the other a teenager who was truant from school. I know about them because I know Donna, who hears these little-known details from Michael. But many people are unaware that the police are pursuing a connection between the two.

There is no mention of these attacks in the local newspaper. The town has its reputation to protect — of being the kind of town where young couples stay, rather than moving to Montreal or Toronto, to raise their children; the kind of town where you don't need to lock your door at night. But even before the murders, women walking on their own have always been frowned on here. There is something unsavoury about it, apparently, as if not having parents to drive you, or who are willing to drive you, ought to seal your fate. Many nights I walk alone to the new library beside the river. The next day in school classmates lean over their desks and whisper, "I saw you *walking* last night."

I drive toward the distant hills and mountains of the north of the province, which stretch away like an inland sea.

More trucks stream by me, their cargo of ex-trees strapped to their backs. Bark falls on the windshield and my nostrils fill with the balsam tang of their wake. A girl in my grade was killed the year before by one of these trucks. Logs the size of houses had been piled on the back of its flatbed, oozing sap onto the chains which bound them. They'd slid free just as the girl passed the truck on the inside of a curve in her Honda Civic.

I watch these trucks warily, my foot hovering over the brake. What would have been the last thing that girl saw on earth? Giant bark rings like the rings of Saturn filling her windshield. Or maybe there wasn't even time for that. Perhaps there was no time for thought at all.

On Sundays my mother drags me to Mass, enraged by the natural reluctance she sees in me for religion. In Mass the priest talks about lives being "snuffed out," like candles. Yes, I decide, it would be like that. A micro-second of panicked perception, a sliver of fear, then oblivion.

"Have you called him?"

"No but I think he followed me."

"Followed you?"

"In his car. He stopped and tried to talk to me."

"Jesus, it could be anyone. It could be that killer."

"Well you don't have any photographs of him, how would I know? It's a great choice, isn't it: get in the car with the father I've never met in my life, or get in the car with a man who will rape and murder me."

"Don't be so melodramatic. All you teenagers, you live at such a pitch I don't know how you don't just explode."

"Well good luck. You've got two more of us to come."

My mother gives me a strange look. Later, I identify the unfamiliar note in it that I'd seen and couldn't place — regret.

The evenings lengthen. I take a third job at the Executive Motor Inn, a hotel, restaurant, and bar complex favoured by the town's big shots — town councillors, police, judges, local captains of industry.

The Exec, as everyone calls it, is located on the shores of the river. Before my shifts I walk down to the river's edge in the smoky dusk. Tables are put out at the back of the Exec bar, so guests can sit overlooking the river. There they are nibbled by the town's languid mosquitoes. They watch the surface of the river, static as plastic.

The river is nearly two kilometres wide. In its centre a current tugs the water toward the coast over 150 kilometres away. Its oily margins are home to June bugs while secretive river birds fly low under the stanchions of the bridge.

I work a shift from six p.m. until four a.m. I don't get into bed until seven in the morning. Before the evening starts, I lean out over the river's edge. The balcony is three or four feet above the surface of the water. Kingfishers fly low over its surface, mirroring themselves so there are two birds, perfect copies of each other.

One day in early May, it is six degrees; the next day it is seventeen. I will never get used to these convulsions, the lack of transition. Even though I come from the adjacent province, I find this province's essential character so different: cloistered, self-regarding. I was brought up by the ocean. This province meanwhile is like a mini-continent. In summer there are three or four weeks of glutinous heat when everyone seeks the salt air of the coast, driving hundreds of kilometres to the sludgy mudflat beaches that hem it.

Now that my escape is in sight, I have gained a perspective on this place where I did not choose to live, where I never would have come if my mother hadn't married her husband. Now I can afford to regard it with a detachment that stops just shy of affection, and which is in fact an abstract kind of regret. I will leave here none the wiser and unchanged for my years in this place. Somehow we have failed to understand each other, this place and I, and it is no ordinary misunderstanding. Already I feel a kind of danger lurking in my final months in this town. I am being presented with a choice too subtle for a seventeen-year-old to grasp. I can embrace or evade its pull. It is up to me.

8. ICEBOUND

néve

Loose granular ice in transition from snow to glacier ice.

DECEMBER 16TH

Sixty-nine degrees south. The world has closed around us. We have come to a stop amid ice floes.

The only fixed visual points are icebergs paralyzed in pack ice. In the far distance are white pulsating smudges — the officers identify these as the mountains of Adelaide Island. Between the floes are thin blue veins. We watch as seals squeeze out of these and flop onto the floe, from where their wet, placid eyes study our behemoth of a machine suddenly stalled in their world, then disappear back through the fissure into a black sea.

The light is scrawny yet intense. Even under cloud it is impossible to look at the ice field with the naked eye. This light has nothing in common with the consoling skies of the temperate latitudes. Now it is tungsten, a slow-burning, blue-white phosphorous. It is broadcast from somewhere inside the brooding stillness of the ice field.

The sun comes out from time to time. To say it stings is not quite enough. It feels like it is burning through us, a cauterizing. Now we are under the force of the spring Antarctic sun. I remember that in spring the ozone level in the Antarctic atmosphere drops by sixty percent, mostly due to the return of polar stratospheric clouds, which cause chemical reactions that diffuse ozone. On the bridge, two drawers beneath the gyro console are stuffed with bottles of SPF 50 sunscreen, which we must wear if we are going outside, even if only for five minutes.

Stuart the deck engineer came on the bridge. He wore a hard hat, engine-room earmuffs, and a soiled white boiler suit with a hi-visibility jacket on top. Spectacles perched on his nose.

As well as being a deck engineer, Stuart revealed he was a trained atmospheric chemist. "I left science because I loved the ocean too much." A furtive guilt passed across his face. "I love living in perpetual motion. I followed my passions."

We retreated to the tea station at the back of the bridge. "Just a little tip," Stuart offered, as I laid out the cups. "The Old Man says he doesn't take sugar but put about a quarter of a teaspoon in."

"What do you think?" I asked, throwing my glance out the bridge windows.

"It looks fast."

With ice, *fast* is not about velocity, but strength — how formed it is.

"Have you seen this before?"

"A few times. Last year." Stuart's earmuffs and his hard hat made him look like a Lego figure: blocks of red and black, piled on top of his beaming *putto* face.

"Do you think we'll get through?"

"Hard to say. It always surprises us. We come South every season expecting there to be no ice left at all down here. But the truth is you can never tell what the ice will do. Sometimes it increases mass."

"But I thought the west Antarctic shelf was disintegrating."

"It is, but that puts more ice in the sea, quite likely. Climate change isn't about how much ice there is in the world, or how cold it gets. It's about climatic systems and climatology, and very few people on the planet understand these, let alone care about them. Possibly even presidents and prime ministers don't. We're rather relying on you to do that."

"I'm not sure I can single-handedly save the situation."

"But that's why you're here, isn't it?"

"That's the official reason."

"Is there a secret reason too?"

"Writing a book is always a secret mission," I said. "Kept secret even from yourself. Because you don't know what you are going to write, or even why, sometimes."

On Stuart's face was a faint note of suspicion, one I would come to see often on people's faces in the Antarctic. Those who do season after season — scientists and ship's officers, communications and IT specialists, directors of divisions of BAS, Foreign Office people and the government ministers — think of writers and artists and journalists as useful to the cause as communicators. But we do not produce knowledge in a hard-facts sense and, like any society, the Antarctic world has its priorities and its hierarchies. The journalists and artists who are put into the continent are witnesses to the scientific findings that are taking place there, but we are not, strictly speaking, necessary.

But then most people, if pressed, would avow that writers, or literature, are not necessary to human society, although they might be a desirable good. While the organization had taken good care of Suzanne and I, I was not sure the scientists knew what to do with us. Journalists communicate in a more direct way: they tend to have an encapsulating type of mind, trained to parse complex information and transform it into the equivalent of an astute phrase, an aphorism. As a writer I lack that kind of mind entirely. I can only see the smoke and mirrors of ordinary experience, and rising out of these necessary obfuscations, in the very far distance, an outline of the truth I must write.

Stuart and I talked over our rapidly cooling cups of tea. Drinks did not stay hot for long on the bridge, which was kept cold to encourage alertness. We went to stare in front of the bridge windows. The chrome ice shield was still there, broadcasting the buried white comet of its light.

The ice fields of the Arctic and the Antarctic are a giant mirror;

the amount of light they fling back into space is called albedo. The Antarctic's reflection rate is about thirty percent, the highest on the planet. The Antarctic is Earth's cooling system, reflecting the heat of the sun back into space. Max's computer model calculated Earth's historical albedo precisely, charting its changes over time as the ice sheets advanced and retreated.

From one of the bridge wing decks, we could see crabeater seals lolling on the ice field below us, thumping and sighing contentedly. Their fur glinted in the sun, making them look like bars of solid gold scattered across the ice field. I imagined a game among hallucinating, deranged Antarctic explorers in which they rushed onto the ice to collect the bars of gold, risking their lives for this great prize, only to find one of the moist-nosed crabeaters snorting at them.

The following day we were still icebound. The *JCR* is an ice-strengthened vessel, but it is not an icebreaker; the difference between the two is horsepower. The ship could slice through ice, but its limited thrust must be cannily employed. Even icebreakers are bested by Antarctic pack ice. It's not uncommon for at least one ship every summer season to spend a couple of weeks stopped by pack — I learned to say this word, *stopped*, or *beset*, and never *stuck*, a serious taboo on a ship.

An announcement went out over the tannoy: two reconnaisance flights from Base R would try to help us find a way out. The pilots of the intercontinental plane, the Dash 7 which flew from Base R to Stanley and back, made a detour on their northbound leg to try to help us.

We watched from the bridge as the pilots approached the ship flying low, propellers churning only a few meters above the surface of the ice, igniting a flurry of snow underneath the fuselage. Apart from Vernadsky, the plane was the first thing that was not us we had seen in nearly three weeks and it was like a beacon from a more advanced civilization.

"Base is only forty kilometres as the crow flies," Martin told me. "It takes them about five minutes to fly out to us, but we might not be able to get to them before January." The year before the same thing had happened: the ship had reached even closer to base, managing to round the southern tip of Adelaide Island before being repelled by pack. "We could practically see the wharf," Martin said.

But the ice had been completely impassible and entrenched. If the *JCR* stayed any longer it would have risked the channel it had broken behind it freezing over, locking it into the ice for weeks. In the end, the ship had to sail all the way back to the Falklands and fly essential cargo and personnel down in relay flights. For the ship to be beset by ice was a once-in-a-decade event; for it to happen two years running was unheard of.

The red plane climbed and buzzed over, not more than a hundred feet above us. Chills travelled up my spine. And also, instant vertigo. I craned my neck and saw sky pans, ice pans, and the plane, a carmine bird of prey.

"I can't believe how low they fly."

"They love it — they're daredevils, all of them," Martin replied. "Watch out when you're on base, they'll have you flying the plane in no time." I thought he was joking, and laughed.

The pilots' radio reports were audible to everyone on the bridge.

The pilots had capable, dashing, slightly militarized voices. They said that the pack looked solid to the south for over thirty kilometres. It was the most disheartening news possible. Even if the wind changed and the ice started to move on the gyre, thirty kilometres of pack wasn't going to shift as quickly as we needed it to.

The plane sliced the air, travelling away from us. I watched it get smaller, until it was a black dot in the sky. In between blurs of static, the captain signed off, "Okay, Greg, thanks for that. Over."

A hiss, then the pilot's voice said, "Good luck. Over."

Sea fog enveloped the ship. Now there were no mountains to puncture the skyline, no visual field, only mist, veils, languid sheets of white muslin. They hung in the air until dispersed by wind, because the light and energy mass balance did not change throughout the day in the springtime Antarctic.

The ship, eerie when not in motion, became stranger still enveloped in this frozen mist. I spent hours sitting in my cabin watching as silver shadows, like rays of ice emitted by the sun, colonized the space.

It was a kind of ambush. On the bridge the officers and seamen had hushed communions; their conversation all about the ice — whether we would get out of it, whether we would make it to base or have to turn around and head north to try later in the season. The bridge windows were like a giant cinema screen. I thought, This is not real. My inability to accept what I saw had something to do with impotence, I think. Rarely in my life have I been less able to affect a situation.

The pack had hardened. From time to time the brown darting shadow of a skua or a storm petrel flashed across the ice. The ice was hard-packed and sliced by sastrugi — sharp, wave-like ridges formed by wind that look like a frozen version of the ridges on an alligator's back.

The captain stared out at the ice field, his face encased in bug-eyed polarized glasses. "Where has all this ice come from?" I asked.

"Pine Island Bay. There's an ice stream there: it's pushing out a lot of ice; glaciers are calving fast."

He led me over to the chart table, where a maritime map of the west coast of Adelaide Island lay, our course marked in a pencilled line of x's. If we went out to sea, he told me, heading to the west, into a lonely body of water with no landfall until Australia, we would very likely encounter the pack again, further to the south where it might be thicker. If we headed east, toward the coast of Adelaide Island, there might be a lead through the ice, but there were also hidden reefs and the charts were unreliable. Unlike the HMS *Endurance*, the Naval polar survey ship and sister vessel to the *JCR*, we didn't have a depth sounder to guide us away from undersea ridges.

"I think we might have to back out," David went on to say. I thought I hadn't heard him correctly. How could we back out when we couldn't even turn the ship around? We were locked in.

"There's another strategy," he added. "We wait, see what the wind will do."

Even if the ice which held us in its grip looked solid, it travelled on the wind and the tide. With the right shift in either or both, it would break up like a puzzle, nudged into disintegration, and we would emerge from this icy maze.

The stalled nights brought out a careening, childish energy. That night the Bar Stool Challenge was won by IT man Bruce, who was from somewhere up the coast north of Aberdeen. The nature of the challenge was to thread your body in and out of the bar stools, snaking through them faster than anyone else. Bruce did it dressed in a kilt, wearing no underwear. We all covered our eyes.

At one point, a man I'd come to privately call the Unfriendly Vehicle Mechanic turned to me.

"What makes you think you can do this?"

"Do what?"

"Write books."

"What makes you think I can't?"

His mouth curled and I saw myself in his eyes: someone along for the ride, someone who would make what sociologists call *symbolic capital* out of this whole experience. I heard it, not for the first time, the unsavoury clang in the phrase, which people on the ship sometimes used when introducing me: *the writer*. This was my role and profession, of course, and no different in its intentions from "the captain," or "the principal scientific investigator," but the term had the odd effect of jolting me out of myself, of making me see myself as others saw me.

What was happening to the writer? She was writing, yes. Every night in her cabin she writes prose poems about ice and desire that she calls "Night Messages." But the writer is spooked, more than ever in her life, by what happened on the ship that night when she thought, if only for a moment, that she might throw herself into the ocean. The writer wants to feel no more anticipation, love, desire; she tries to turn these off at source, but like a dormant computer programme they are there, humming away insistently, underneath her heartbeat.

Meanwhile her mind is a trap made of words. All her life, words have been a talent, yes, but even more than that an exile, a consolation prize.

She has been so ecstatic on this journey, released into knowledge and discovery. But also miserable, incarcerated, addled. She has never before felt this particular helix of emotion, or even thought it possible. Her life before these days was a shadow, a waning memory of a dream.

The Antarctic has always required people to identify what they value, what they are prepared to risk and to lose. Even so, I don't know why this all felt so final in this place, on this journey. Something about it seemed bent on determining my life, both my past and my future. I felt the sense of a snare, of a destiny not formed by my own instincts and desires, but by some intent located in the boxy unfamiliar constellations in the southern hemisphere sky.

This journey by ship would resound in my life for many years. I didn't know yet how it would cast a long white shadow over my heart.

"The stock and the till take just aren't adding up some days."

Elin frowns. Long amber shadows fall between the tallow candles, the coasters crocheted by old women in hamlets by the river.

Elin is tall and what people used to call "big boned." She has blonde hair bluntly cut and a long face.

"I don't know why that is, I'm sorry." The adult cadence to my voice is new. It is something I have only recently acquired. I like trying it out.

"Let's go over the list."

Together we pore over sales of earrings, sand dollars, gift cards, wrapping paper, stained-glass ornaments, crystals dangling on transparent plastic string, stuffed lobsters and seals.

Eight dollars is missing, plus sales tax. I pretend to peer at the numbers column but I am looking at Elin's shoes. She wears flat leather wraparound sandals I have never seen before with a name emblazoned on their side: *Birkenstocks*. In the winter Elin wears leather shoes which look like the shoes worn by hunch-backed medieval peasants in my History of Europe textbook. "Granola shoes," says Donna when I tell her about Elin's apparel, wrinkling her nose, meaning hippy.

"Can you explain that discrepancy?" Elin has a Scandinavian accent that gives her voice an authoritative edge. She comes from a fair-minded country. Denmark, or Sweden. She will give me a chance.

"I remember entering the earrings as $10, but is it possible they were $18," I say. "An eight can look like a ten."

"I need you to pay careful attention to the price tags. No matter how much you're chatting to customers" — this was another thing Elin said: *chatting*, not *talking*.

It dawns on me that she suspects me of something, other than fatigue. This has never been a trusting town. It is the kind of place where people worry they are not getting the best possible deal out of life and begin to cast looks at their neighbour, wondering who is to blame. The most extreme form of this might be the murderer, whoever he is.

"I will," I say. "Promise."

Apart from the woodlot and the river and the library, the only places to go in town are stores. Stores are refuges, or would be, if the sales assistants' eyes did not snag on you, willing you to buy. Stores sell neon clothes, scented bags of potpourri, the snakeskin pumps all the girls in high school are wearing that year.

My favourite is the Canadian Tire hardware store. I go in to get away from the heat or cold, depending on the season. I find peace wandering the aisles that smell of rubber and steel. A manly, useful smell, cooled by the chromium air pumped out from the ceiling.

Men in red shirts ask if they can help, if I am looking for anything in particular. I make things up: I am looking for a sink plug, a stovetop reducer, things I only know exist thanks to my job at the Exec and because I am forced to care for the children my mother and Mark have now. I linger in the blow-up paddle pool section, dazzled by the blue water and the unreal green of the pools, comparing weight and volume, the time they take to inflate.

The days extend themselves with a reluctant sinuous languor. Now it is light until eight in the evening. Blue hills stalk the river valley. In the distance, the hoods of mountains. Low and exhausted, these are the last of the Appalachians. A hollow yearning, completely new to me, installs itself near my lungs. I learn to carry this new hollowness around with me like a phantom limb; before I had been complete.

There is nothing I will miss about this town, other than my friends. I know its rhythm now, the spring and fall exhibitions when farmers bring their livestock for show, the purple world of the fairground rides and the rigged games booths that can be heard from blocks away, the long summer days spent at the beach on the man-made lake by the dam.

All night I serve food to the people this town has made, although it is the alcohol that people come for; the Exec has a late alcohol licence. The men who come to eat Chinese food at three a.m. in the dining room are of a piece, sired by the town and the fertile river valleys that surround it — chunky, check-shirted and cowboy-booted, blunt bodies, their faces a mix of Scots, Irish, English, Huguenot. The occasional upturned eyes and narrow, elegant nose that bespeak Native blood. It is these men I like: slim, delicately curved, like the stems of flowers.

It is four, five, six in the morning when I finish my shifts. I don't risk walking. I drive or get a taxi home. But I still walk home from Shades of Light. I walk to the library every night too.

Everyone I know cautions me to stop this. Even Donna's mother, who disapproves of me, tells me to take better care, that she'd hate to find out something had happened to me.

"What can you do?" is my mother's response, when I tell her about my lone walks, her eye on the baby monitor. "You can't live your life swaddled in cotton. Eventually bad things find us all."

9. EROS THE BITTERSWEET

stamukha

A single fragment of ice stranded on a shoal.

DECEMBER 19TH

We have been stopped for four days now. We stand on the bridge in the wraparound BAS-issue sunglasses that make us look like Steven Spielberg aliens, staring at the ice. We just want it to let us go.

Now, everything we say, every chance meeting in the corridor carries the charge of a live wire, like the electrical shocks that plague us. The ship's four engines are turned off. In the hollow silence I imagine sounds — unfamiliar antique church bells, the clangs of farriers, the sound of knives sharpening.

Max visited me in my cabin, his eyes migrating compulsively to the window, into the glare of the ice field and the stalled rapture it inflicted upon us.

"I don't know why I'm feeling all this now," he said. "I can't get a handle on my emotions. I don't like being out of control."

"It's not possible to be in control all the time."

"I know, but I can't tell whether this is a damaging experience, or a good one. For the first time in my life, I can't make that judgment."

"Everything that we feel, or see, or experience leaves its imprint upon us. You have to be careful what and who you let yourself get close to." I was impatient with myself. Why was I offering him this grim chalice of advice?

"You sound like you don't trust people."

"It's circumstance I don't trust. Life is volatile, much more so than people think, and almost completely beyond your control. You have to approach it like a ship sailing through ice-infested waters."

He gave a smile then — the first I had seen in some days. There was tolerance in it, the kind of patient amusement you might bestow on a parent.

"Here, geometry always makes me feel better." He sketched isosceles triangles in his swift, certain physicist's hand. "Look, this is the rate of sublimation of the ice sheet, remember I told you? This is the subduction zone."

I misheard him and for a second thought he'd said *seduction* zone. Sublimation, subduction — it was an emotional tongue, this ice language.

He opened his laptop and I could see again see those streams of numbers and letters separated by coded prompts. It amazed me that from these sequences of numbers and letters he could concoct vast sheets of ice that have not existed for 100,000 years.

Max told me how ice crystals' exact shape depends on temperature and vapour tension. Prismatic crystals are short, solid, or hollow; needled crystals are almost cylindrical. Plate crystals are more or less hexagonal, he said. The star crystals are, as you would expect, stars. Then there are the frosted particles, or the clusters of tiny crystals. Glaciers and icebergs are made up of innumerable crystals of these varying designs, each of them in internal motion.

"But how does a glacier begin life?"

"As snow, freezing rain, ice condensed directly from vapour, freezing meltwater, avalanches. Snow, crystals, trapped air, tough ice called firn, which hardens into glacial ice."

"*Firn.*" It had the sound of good whiskey, or a task.

He scribbled an equation about mass and velocity. "Look, there is a velocity, a net loss and gain of mass. There is ablation, melting, the flow downhill into the sea. In the western Antarctic the ice sheets rest on bedrock that is already below sea level, so the downward trend is reinforced."

He paused to draw another swift hieroglyph. "I'll show you how a glacier moves. There are several ways — by kinetic waves, which are something like a flood wave, like a frozen tsunami."

I thought of the watercolour by the great Japanese artist Katsushika Hokusai, the arcing, giant wave, its arresting mixture of movement and classical restraint.

"Mostly by basal sliding — that's when the ice is lubricated at bottom, where it touches rock, by the heat generated as the weight of the ice meets the resistance of rock. Glaciers can also flow upward, to an extent, in ablation zones, and true to their name, cirque glaciers seem to move in a circle. But glaciers are never a single ice terrane, they're several scattered ice masses."

Max reeled all this off as if he were talking about a journey to the supermarket. It was internalized knowledge; he'd been studying glaciology for the last three years. I allowed the logic and process of his explanation to slide by and again fixated on the language he used, its medicinal tang, *terrane, cirque, tongue, ablation,* how it commuted between anatomy and architecture.

Later, we hung over the rail of the bridge, looking down on all this ice, so blank to the untrained eye but, like any desert, actually bristling with detail once your eye has been made alive to it. But in the frozen détente between our ship and the ice our predicament was worsening. We were down to a final twenty-four hours; soon the captain would have to turn the ship around and we would steam back to the Falkland Islands.

Or, if we could reach base, we would be there within a day. Many people on board, Max included, were booked on a flight to the Falklands, then to the UK on December 23. Everyone wanted to get home for Christmas. Base was tantalizingly close, the kind of distance you would traverse on a train in half an hour, yet it was unreachable.

For two days we had been sallying to prise open the grip of the ice. Sallying is a deliberate see-sawing of the ship from side to side using a release of thrust from steam built up in the engine valves. In the old days on the explorer ships, sallying meant getting everyone on board, including the dogs, to run in a gang from one side of the ship to the other. We were now accustomed to the roar and sigh as the ship heaved, lolling several degrees to the right, then the left.

When the sallying failed, Captain David turned off the remaining engine. He entered the dining room with the look of a doctor who had tried defibrillation, adrenaline — everything — to restart a heart. He came down to the bar and sat on the banquette, arms folded across his chest.

"How do you feel about spending Christmas on board?"

"That depends," I said. "Have we got any turkey?"

No one laughed. To our meals and our board games there was now an unmistakeable sludge of claustrophobia. Another day in the ice meant that Andy the engineer would not get his trip to the South Pole to fix low-power magnetometers. Another day in the ice meant that I might never see the base myself. A writer was hardly essential personnel and the organization might not fly me in, should we have to turn back and go to the Chilean base, Marsh, where there was an airstrip. I might be put on the next plane back to the Falklands, rather than Antarctica. So close, but yet so far. It had never occurred to me that we could be stopped a mere sixty kilometres short of landfall.

I went outside on deck again to survey our prison. The contented crabeaters were still there, lounging on the ice just below us, snorting and sighing. This is the Antarctic, I said to myself. This is the place where Mawson was pulled alive out of innumerable crevasse falls. Where Shackleton took four months to rescue his stranded men on Elephant Island. Scott and his men, who died for eighteen kilometres and a badly placed depot. And the American aviator Richard Byrd, with his shipment of straitjackets to the continent, lest his men lose it in the long winter. Why, after reading all the classic explorer narratives, with their catalogues of misfortune and near-misses, of logistical disasters and bizarre rescues, did I think it would all go smoothly?

That afternoon a lead finally broke through the ice, but it turned out to be a blind lead — a channel of water that leads only to another white tongue of ice. After a couple of hours of creeping through it at four knots, we ground to a halt.

The sun had come out and colours were reversed: a black sun, black ice, and white sky, white water. I couldn't remember how many days we had gone without night, now. Before this trip I had never lived a day of my life without seeing night. Could it be that my eye was manufacturing it, creating dark where there was only light?

That night we had a Christmas decorating party. We vented our frustrations on the Christmas tree, choking it in ugly garlands, and on the alcohol stock behind the bar. We drank everything in sight while Bruce the IT man undertook another Bar Stool Challenge in his kilt, scattering us to the far side of the room.

Max and I did not seek each other out amidst the raucousness. A very subtle note of rebuff had crept into our conversations. He may well have picked up that I had become attached to him, and he had been intending to abuse my sympathies by exiling me, but he held back. That was one limit of his compassion. But on the ship it was hard to tell what was really going on between people. The confinement forced those of incompatible temperaments to live as a family, sharing meals and chores and work and DVD screenings and nights in the bar for hours, or it threatened to erode natural sympathies, as between Max and I, with the boredom that comes with overexposure.

After the decorating party, I retreated to my cabin. I had to do this many times in those days. I was so used to spending my days reading and writing and had become unaccustomed to the blare of circumstance, to the banter and jokes that are the currency of group dynamics.

In my cabin, I opened one of the few non-polar books I had carted with me: Anne Carson's essays *Eros the Bittersweet*. Carson is a poet and a Classics scholar, also a renowned translator of Sappho. In ancient Greek, she explains, error (*hamartia*) was something which had to be gone through, in order for life to progress. Aristotle put it more bleakly — error is the false step that leads the protagonist to his or her tragedy, but in meeting that tragedy, the hero was set free. Errors have a larger purpose which we cannot, in the moment of committing them, determine.

Sappho wrote that love was "sweetbitter." The word, *glukupikron*, Carson explains, puts the sweet before the sour. The word sounds different with the senses swapped: first the sweet, then the bitter. Desire has a language of heat, of liquidity and melting, Carson notes. Sophocles compared the experience of desire to a lump of ice melting in warm hands, pleasure and pain intermingled: the bite of cold, not kind, built from ferocity, then the hot liquid charge of melt.

Love, Sophocles said, takes shape through a series of crises of the senses. A *krisis* calls for decision and action. There is a blind spot in Eros, a moment of time in which absence and presence converge to make a slim paradox. You bite into the long-desired apple, and await the sweetbitter moment when your desire will be gone.

At the centre of desire, Carson writes, is a cold, primal pleasure. Ice is pain, but also a novel enjoyment. As a child, a snowflake on your tongue or face gives you a shock, and this shock spirals upward, throughout life, into the logic of your existence, into the psychology of desire, and the shock wanes to the point that as adults we can withstand an entire blizzard and not feel anything.

Ice is a novelty, then, still, and like love, an unexpected form of torture — fresh, untried. But then there is also the catastrophe of its dissolution: the longer you hold it, the colder your hands get. The longer you hold it, the faster it melts. Carson seems to take this twinning of hot and cold, of fire and ice, as a way to explain the trauma that love enacts on the heart.

But cold also burns. Look at any picture of frostbitten fingers and you will see they are charred black, as if they have been held to a flame.

Hamartia, *krisis*, *glukupikron*, *cryos*. Error, crisis, bittersweet, ice. That so-called night trapped in the ice, I took a strange consolation in this cool language.

I went back upstairs to the bar to get a bottle of water. I was tired of the flat, ionized taste of the water the *JCR* so capably made in her desalination tanks. I found a party in full flow. Someone had drawn the blinds against the blaring sun and put on a Christmas carol CD at full blast. We women were corralled to sit on the bar stools to be decorated: Santa hats on our heads, garlands of tinsel around our necks. Mike, the purser and the man with the keys to the alcohol store, brought up extra supplies. "Yez can all drink as much as yez like."

We danced to "Rockin' around the Christmas Tree" as a last-ditch sally was about to take place. As the ship rolled, its engines exploding steam in an effort to unlock the ice, we lurched heftily to starboard, then to port. We started on scotch, on rum, on beer, on wine, and ended on angostura bitters with whatever was left.

In the morning, we would turn around and head north, shunned from arriving at Base R by the greater power of nature. I would remain in the Antarctic for the three days that it would take us to steam up to Marsh base of course, but the sense of discovery, of making a journey which I had never made before, would be gone.

I couldn't sleep that night. I began reading another of the books I had brought, Fridtjof Nansen's *Farthest North*, an account of his overwintering expedition to the Arctic, locked fast in the ice. Like his brethren, Nansen was a gifted writer, moved by his environment to an awed precision for description.

It was five in the morning, and the bright tangerine sun of the Antarctic coated my cabin. I finally fell asleep on the page that ends with this passage:

Nothing more wonderfully beautiful can exist than the Arctic night. It is dreamland, painted in the imagination's most delicate tints; it is colour etherealized. One shade melts into the other, so that you cannot tell where one ends and the other begins, and yet they are all there. No forms — it is all faint, dreamy colour music . . . The sky is like an enormous cupola, blue at the zenith, shading down into green, and then into lilac and violet at the edges. Over the ice-fields there are cold violet-blue shadows, with lighter pink tints where a ridge here and there catches the last reflection of the vanished day . . .

I have never been able to grasp the fact that this earth will some day be spent and desolate and empty. To what end, in that case, all this beauty, with not a creature to rejoice in it? Now I begin to divine it. This is the coming earth — here are beauty and death. But to what purpose? Ah, what is the purpose of all these spheres? Read the answer, if you can, in the starry blue firmament.

PART TWO
MEAN SOLAR DAYS

1. PRECESSION

Fast ice

*Consolidated solid ice attached to the shore, to an ice wall
or to an ice front.*

DECEMBER 20TH

We are all outside on deck; we mill about in excited knots, our eyes shut behind our sunglasses. We sail past the black flanks of Jenny Island. Beside us the dark dorsals of minke whales. We struggle to accept the change in our reality. Only a few hours ago we were caught in a fist of ice and now we are doing sixteen knots, accompanied by this wing-guard of minkes.

Somewhere on the other side of a horn-like headland lies base. On the bridge it looks so close. The readout shows three or four rectangular lozenges — the base buildings — and a long brown strip bisecting the peninsula — the runway.

I hadn't realized how much I missed the sound of the engines, their reassuring rattle. It feels like a miracle, simply to be moving through the water again, which, unfrozen, returns to being a conduit, a friendly substance.

Veils of mist hang low to the surface of the ocean. The sea is a deep tanzanite blue. We round the headland, listening to whales breathe alongside us.

Early that morning, around six o'clock, Captain David had found an exit from the ice maze. He decided to chance it and travel east, toward Adelaide Island and its bony reefs, wagering that the pack would not blow in and trap us there. For four hours we crept along the reef at a slow speed. If we hit one, even at six knots, the

JCR's reinforced hull would be rent open like a can opener slicing through a tin.

I woke later that morning after only three hours' sleep to find the ship in motion for the first time in four days, slinking along Adelaide Island, the mountains of its western spine glowing blue-white in the morning. Skuas began to appear once again. Skuas are land-based, in contrast to the albatross, which can go for many months without putting their feet on solid land. A skua at sea means you are close to land.

The sun glanced on brilliant blue open water. Wisps of cloud hung between the mountains, so low they hovered only a few feet above the sea. We passed through this veil of vapour clouds, which remained intact in our wake.

We were all out on deck for the arrival. On base we could see stick figures at the top of a hill of granite stone, waving. The ship's arrival was as keenly anticipated by those on base as it was for us: we carried the summer's supplies, including innumerable crates of beer.

The *JCR* sidled up to the wharf and the familiar churn of the ship's engines receded. New sounds took its place: the distant buzz of a generator, wind zinging off wires. Around the back of the base was a tangle of radio transmitters and satellite domes. The VHF radio transmitter was strung in a crazy trapezoid-oval, a dreamcatcher ranging northeast, toward England.

I was surprised to find I didn't like being against solid land. I didn't want to leave the ship. It was my first morning in the Antarctic, the mysterious continent where only a few lucky humans ever set foot, and I had to force myself to take account of my surroundings.

From other people's reactions I knew I wasn't alone. Caroline the diver and the Unfriendly Vehicle Mechanic were looking at what would be home for the next eighteen months, or even the next two years, should they decide to extend their contracts. As we gathered in the ship's bar for the Base R base commander's briefing, we were all uncommonly subdued.

"You will disembark the ship at ten hundred hours," Simon the base commander informed us. Suddenly I no longer liked being told what to do, although I had accepted it while at sea.

But we had no choice, because the ship no longer tolerated us. We were told to strip our beds. "Anyone who doesn't won't get a fockin' drop to drink on this ship from now on," Mike the purser warned. "And remember to leave your bedding outside in the corridor."

And suddenly that was it. Mike settled our bar accounts and handed us back our passports. Max, Nils, and Emilia would remain aboard the ship until they flew out in three days' time. I was jealous that they could remain on the ship, that they didn't have to commit to the new human ecology of the terrestrial base.

As if on cue, a roar filled our ears and the Dash 7 appeared, its four turboprops grinding the sky, its red fuselage framed in the windows of the bar. At the last minute the wing slipped by the ship's conning tower and disappeared behind a cloud of runway gravel dust.

"Chancers," muttered the purser.

Almost immediately Simon gathered those who would be staying on base for a walking tour of our new home. As we stood waiting there was a sullen note in our postures; a few hours before we were desperate to arrive at base, but now that we were there we recoiled. Instead of staring in wonder at its impressive position, we dropped our gazes to the tips of our rigger boots.

My first impression was of sounds. I had become used to the tinny silence of the ship stopped in ice, the only sounds the grind and wrench of the sea ice as it nudged the hull. Base emitted a hushed grandeur, as if a vault door had been closed, and all sound guarded fiercely behind it. A gunshot crack Dopplered through the air from time to time — icebergs imploding, I would come to learn, or rotating, and a sound like a distant waterfall when a piece of the ice shelf collapsed into the bay. Antarctic birds are largely soundless, but sometimes we heard the rough squawk of a skua, magnified in the empty air.

It felt warm and cold at once. The air was dry, as if we were inhaling paper. Meltwater rivulets coursed alongside the paths but it lacked the dewy, ionized smell of water. There were no smells of soil; no trees nor grass nor flowers. I did perceive a smell that I would come to associate with the Antarctic and that I would never experience again, except to a muted degree in Greenland, but if I were to smell it now, I would recognize it immediately. When asked what the moon smelled like, astronaut Buzz Aldrin, the second man to walk on its surface, said, "Two stones rubbed together." That's the smell of the Antarctic — flint.

We passed men in orange padded boiler suits driving JCBs. The jetty apron was frantic with activity — containers swung, snagged by cranes. The Dash 7 appeared again, this time on a takeoff run, roaring into the air. Rugged men walked toward us wearing slit-eyed sunglasses. It looked like the set of a polar James Bond film.

We rounded a corner and were confronted with a series of

boxes on rocks. The largest of these was the aircraft hangar. It housed the Dash 7 and two Twin Otters when they were at the base.

We threw each inscrutable other looks which were reflected back to us by our mirrored sunglasses. We had all been briefed, of course: how Base R's role as an Antarctic airport was key for BAS' operations. How it was the largest of the British bases and in high summer, January and February, up to a hundred people could be accommodated. Many were scientists and field assistants returning from tent field camps or being deployed to them. This, as well as the presence of "drop-in" visitors (anyone staying for less than a three-month tour was considered a fly-by-night at Base R) such as Foreign and Commonwealth Office lawyers, BBC journalists, and government VIPs, gave the base a transient, cosmopolitan feel, we were told. Still, the granite quarry-crossed-with-a penal-colony air of the place unnerved us, accustomed as we were by that point to three-course dinners on the ship, and to the gracious, purposeful ethos of life in motion.

We were shown the boat house, the laboratory, the dining hall / office complex, which also harboured the communications and air traffic control tower and the pilots' offices, the accommodation complexes, the food stores, the "sledge store" where the skis and mountaineering equipment were held and repaired, the skidoo garage, the sewage treatment plant, the generator hut, the Miracle Span — a machine that gobbled everything and turned it into neat bales of waste, to be removed by ship. And that, bar a couple of small storage buildings, was it.

Simon left us to wander around base. We peered in clouded windows to see large packages wrapped in silver paper or ancient skis stacked like lumber against walls. It all looked like a giant Joseph Beuys installation. It felt exciting, in a forlorn, abandoned way, to know that this shed town would be my home for months to come.

But there was an element of siege in the place too. There would be nowhere to go for months, other than the collection of huts we had just passed through. No parks, no art galleries or museums, no cafés or bars or restaurants or seminar rooms or lecture halls, no trains or planes or shops or supermarkets, hairdressers, clubs, pubs, beaches, or swimming pools. No animals, or children. And crucially, no strangers, past a certain point at least. At no moment would someone not know where we were on base; safety regulations forbade us going off on our own. Privacy would be difficult to come by. We were voluntary exiles in a reduced civilizational plane not too different from a medieval village. Perhaps we hoped our time here would teach us something necessary and irreplaceable about the world, would furnish us with an unordinary knowledge which our previous life, rife with distractions and bombarded with trivia, had denied us.

At lunchtime Simon led us to the dining hall. We were told we had to first remove our rigger boots in the boot room. Battered boots the colour of butterscotch scrawled with the owners' initials had already claimed the benches. Ours, newly minted, were the tan-orange of fresh caramel. Hi-vis jackets streaked with grease hung on the nails.

We left our suspiciously new clobber behind and padded upstairs in our socks. As we queued for food in the cafeteria those of us from the ship talked among ourselves. The base personnel wove in and out of each other, cajoling, teasing, slapping each other on the back. They did not interact with us or acknowledge the presence of newcomers.

Next we were lined up against the filing cabinet in the doctor's surgery and Melissa the base doctor — who, I would discover, doubled as the photographer and the hairdresser — took our mug shot. This picture was to remain on the bulletin board outside the base commander's office for the duration of our stay. We were given a little plastic bar with our name typed on a piece of paper, which was then taped to the bar. The plastic bar had a small hole on one end, for hanging on a nail. This, we learned, was the most essential piece of equipment for our stay — our tag for the tagging board. For the next weeks and months we would be required to record our every move. The tagging board would tell whether we were *On Base*, *Off Base*, *In the Field*, *Flying*, *On the Runway*, *In the Hangar*, *In the Bonner Lab*, *The Boatshed*, *Local Travel Area*, or *Around the Point*.

At eleven that night our group from the ship decided to go skiing. I couldn't ski so I drove the skidoo up and down, towing people up the slope. On the way down to base I stopped halfway and cut the engine. Rob, a field assistant whom I'd met briefly earlier that day, pulled up beside me. "Quite the view, isn't it?"

In silence we both looked at the white ring of mountains. The sea ice that plagued our passage down had drifted into the bay. The ship was snug next to the wharf below us, locked in ice once more.

The silence, while convincingly silent, was missing something. I thought, soon I will hear the wash and draw of the tide, the slide of water over rocks. Soon I will hear the swish and hum of the slim wake of a boat, or the shout of a human voice.

Then I began to hear barking dogs, tinny music, people shouting.

"What's that?" I asked Rob.

He stared at me, perplexed. "What?"

"I thought I heard something."

I tried to shake the sounds out of my head, but they became more insistent. I could hear the grind of a Ferris wheel, the clink of money cascading through a machine. A ghost carnival had installed itself in my mind, in defiance of so much silence.

I shivered. Rob's gloved hand lingered near the Skidoo ignition. We'd been standing still for too long. I was trying to hold it at bay, what the Inuit call *ilira* — the mixture of fear and awe that landscape can provoke. I thought, I will never be able to write about this place.

Suddenly I was cold. My hands were encased in black padded skidoo gloves and I wrapped them around my shoulders. My eye was drawn again to the hulk of Jenny Island. In the waning midnight sun, it was impressive, its battlement rising rigid from the sea. In between its two bruise-hued flanks a glacierlet lolled. The island had a wolfish quality.

That night sitting on a cooling skidoo I couldn't know how so many times in the months to come I would look at this island more than any other feature in our landscape. How it would become jailor, oracle, fortress of the dead.

DECEMBER 21ST

The southern hemisphere solstice. Today is the longest day of the year; the sun will hover high over base, the lazy ellipsis it will draw the only sign of its passage between day and night.

I stand on the veranda looking out toward the wharf. I have never seen colours like this before: a thin gold coats the pancake ice that has formed in the bay. The sky is olive and around the sun is a mango gleam, fluted by flares of an intense, fluorescent white. This is what it might be like to stand on another planet: to see the sun from the light face of a nightless star, somewhere on the other side of the moon.

There was a party on the ship the following night to say goodbye to the group who would fly out the next day and also to the *JCR* itself, which would leave that same afternoon.

Somehow I hadn't registered that the officers had come all this way, struggling to arrive here, only to leave after thirty-six hours. But that was Antarctic life: people come and go and the ship had work to do, a survey down near Pine Island Bay. As I would learn in the coming months, the same rule of human engagement that applies in normal life is also found in the Antarctic, but with more intensity and consequence: the people with whom we feel a natural felicity are usually on their way out before you know it, while the people you struggle to connect with stay until their presence is a kind of taunt.

The ship's hold was empty. The JCB construction vehicles and drums of aviation fuel, known in the Antarctic world as avtur, had been unloaded. We had spent nearly two days carrying boxes in a human chain, loading the kitchen stores, alcohol stores, the freezers and the sledge store, carting ski wax and sledge rigging made of reindeer gut, as well as less exotic items: enough boxes of tinned mushrooms to last a century, surely, and cases and cases of Guinness-in-the-can. In Antarctic-speak, unloading the ship was called *relief*. There was no such relief for us, who had aching arm muscles and split nails from the cold and the heft of so many boxes.

For the party I returned to the ship, so recently my home. I couldn't stop myself from passing by my cabin, but it was no longer mine. A hard hat was placed proprietorially on my bunk.

Mike the purser came barelling down the corridor. "Couldn't stay away, could you?"

I watched Mike walk down the corridor, his put-upon gait, his shiny Clarks brogues now so familiar. I could stow away under a bunk in an unoccupied cabin. Nobody would notice, would they? I could go for five days without food, or I could sneak into the duty mess at four in the morning and survive on toast and cheese. We'd be back in the Falklands before anyone would raise the alarm and at least I would have lost some of the weight I put on eating cooked breakfasts every morning.

I didn't know why I had lost my nerve so quickly. But even after only a day on base, I knew we were in a new dimension, that the Antarctic is extraterrestrial in that it has no knowledge of humanity. Humans have never lived there and so the land does not remember us. I found it strangely pleasing to have been so forgotten, but I wasn't sure I could withstand this amnesia long-term.

The party in the bar ebbed and flowed with people from base. The base population was entirely made up of fit young people sporting a telltale Antarctic panda tan: their faces a brick sunburn and a white stripe where their sunglasses covered their eyes. Us ship dwellers looked meek compared to these hale new people. A fume of the outside world clung to us still, as if we might change at any moment into skinny jeans and leather jackets. The base people wore the exact same clothes: a fleece the colour of blood oranges and green moleskin trousers inside, orange padded boiler suits outside. With hats and sunglasses donned, it could be quite difficult to tell one person from another.

"We've been going to the same bar for a year," Rob, a field assistant, said to explain the base people's apparent euphoria. I remembered what Elliott had told me about base life on the ship, only two weeks before: "It's like a cross between university and the Army." They did look as if they'd been through something together, all these ruddy, vital people. They also trasmitted a hunger, an impatient, almost manic appetite to talk: "fresh meat," perhaps, as Max had said, noting the ravenous yet resentful note in the Falkland Islanders' gaze.

Eventually the base people returned to their pit rooms and by three in the morning it was us, the ship people, who were left to drink the bar dry, to be a group for the last time.

Max and I didn't speak. He was buoyant with new acquaintances, but aloof and alert with me. There was an irony in all this I couldn't quite define. Max had been my accompanier on this journey, but he was the only person I couldn't speak to, now that it was over. His rebuff was as plain as if it had a physical prescence; he had his own Antarctic Convergence ringed around him. How easy it is to end up in the company of people who we could take or leave, who are pleasant and kind, but how difficult to find someone who is like another, possible yet impossible, version of yourself, and you them. And when we find this strange affinity we can't resist the temptation to abandon it.

I went outside to observe the light. The sun was suspended above the horizon, bathing the ice floes in Ryder Bay in a cold peach gleam. The clouds — stratus and altostratus, high-altitude, slim clouds — emitted lemon prongs of light. A flute of copper was flung across the ice field in an unhinged flash; the ice turned rose. The colours changed by the second; now lavender, now mint.

I wondered if I would ever be able to get a perspective on this place. I could see that we had all — Max and I especially — been thrown together in that great human lottery known as the Antarctic. There, we might be temporary friends with a pilot, a plumber, a marine biologist, a radio technician, a writer, someone ten or twenty years our senior or junior. There, freed from our little cordons of demographics and peer groups and shared interests and activities, it was possible to meet anyone. It was a social experiment, a frozen Ark. But this capsule of random affinities would by necessity dissolve once you were no longer in Antarctica.

I had promised to give Max a present when he left Antarctica. I left it on his desk while he was up at the glacier enjoying his final ski run. My gift was a weak parody of Edwardian Antarctic explorer diaries:

September 9th

One pony — poor Trixie — lame. Not much longer to go, I fear. None of us relishes the thought of pony stew.

September 10th

Sledged all day then, two feet from camp, all fell into a crevasse. Dogs dangled by harnesses thirty feet deep, sledge arrested fall. Pulled ourselves out, using ice axe and ropes. But all forgotten in the warm bosom of a jolly good cup of tea!

September 12th

All but one of us afflicted by snow blindness. Better not to see where we're heading, I wager! Poor Tiny went over the cliff edge today.

September 13th

Ate the last dog — poor Boris. A bit tough, but better than penguin leg.

September 15th

Everybody dead, but in good spirits.

In the morning ten people stood at the edge of the veranda, waiting to follow Simon across the runway to the hangar, where the Dash 7 awaited. Among the departing personnel were Nils, Emilia, and Max. I caught a glimpse of Max among the crowd. He was laughing, his head thrown back, his mouth open, thin muscles straining on his neck. He was keen to get home for Christmas, I knew, to join his mother who was waiting for him in Zurich. In twenty-four hours he would be there.

Max left me two presents. I found them on my desk in my new office, Lab 7, after the plane was gone. A clock and a book, *Introduction to Geophysics*, sat next to my laptop accompanied by a long note written on pink Post-it notes. He wrote a few lines about the book, and why I needed the clock ("to help you keep track of time!"). At the bottom he drew a smiley face underneath his name.

I watched the plane take off from my office. The Dash 7 taxied to the end of the runway, its propellers grinding the air. At its end, where the runway met the bay, the plane performed a deft balletic turn and stood for a few seconds, its frame trembling as the propellers' thrust was increased. Then it tore down the runway and was snagged into the air as if by a wire. I craned my neck to watch as the plane was eaten by sky.

Max's departure instantly emptied the Antarctic. I felt an indefinite hollowness, but also relief. Some sort of ordeal had ended with his leaving.

At midnight I went for a run on the sunlit runway. Running in that brand of cold was a strange sensation — hot on the inside, while the stinging cold of a thirty-knot wind penetrated the heat and cooled it, but irregularly, in stripes, so that my body felt carved into separate strata. Later I grabbed a cup of hot chocolate in the deserted dining room, went to my office, and read.

The geophysics book Max left me had a chapter on precession, which means the movement of the rotational axis of a spinning body, such as a planet. In the case of our planet, the earth's rotation very slowly produces a cone-shaped graph, although it takes 26,000 years of spinning to produce the completed shape. The exact angle, or inclination, of the earth's orbit drifts up and down the axis. *Precession of a spinning body occurs when it experiences a torque normal to the axis of its rotation*: this was Max's language, the tart, measurable cant I would make my bedfellow in the Antarctic.

I realized I was ignorant of even the most basic workings of the planet. It's so easy to forget that we live on a moving sphere. Earth is spinning at a rate of 1,300 kilometres an hour at the equator and it spins hardly at all at the poles. A thousand miles to the south of Base R, at the South Pole, the earth rotates approximately a centimetre an hour. How is it that equatorial people don't get horizontal vertigo? And here, where I sat at the foot of a glacier on the Antarctic peninsula, did the world feel slower? Did we feel the stasis, the still point of this turning world?

The centre of the earth is molten, like an interior sun, and the earth's magnetic alignment is not stable but wavers between north and south, trying to choose its horizon. We are unaware of these energies, largely, while whales and terns register its fluxes with their sophisticated navigational brains. For the first time in many years, possibly since I had been a child, I felt a basic wonder about the planet and its workings.

I put down *Introduction to Geophysics*, suddenly aware of the time. Base had gone quiet; no footsteps creaked through the ceiling of my office, no ragged bursts of laughter came from the bar. The sun beamed in through my window. The clock said it was midnight, but my watch said two thirty in the morning. The clock Max found for me had stopped ticking.

The murderer's third attack does not succeed. It was a lone woman jogger on a rare rainy evening, running by the river. Michael tells Donna, who tells me. The woman fought for her life. She was very fit; she'd done judo, tae kwon do. She didn't get a good look at him, Michael says; his partner Paul interviewed her. "It was like she was in a daze," he said.

The incident does not appear in the newspaper. But everyone talks about it: at school; in the downtown café, La Vie en Rose; in the queues for the tellers in the Bank of Montreal. But in the newspapers and on the radio and from the police there is silence.

"They don't want to ruin the reputation of the town. Nice place to live, nice place to grow old and die. They're happy to let a couple of young women die so the Tourist Board can go about its business." This is Elin's take on it. Elin can say such things because she is Scandinavian and they are opinionated there.

People in that town do voice opinions, about whether to buy a Ford Fairmont or a Taurus, that new refrigerator, that insurance plan, the state of the roads — how many potholes are left over from winter. Or the attacks on the reservation, white boys and reservation boys shattering each other's windshields with baseball bats. There, boys with eyes like mussel shells play hard-drinking games of pool in smoky bars.

Donna and I are in her house, her parents are out. Michael occupies another corner of their gigantic rec room, lifting weights.

"Why do you think he does it?"

Michael wears only a T-shirt and shorts, even though the night is cool. He is doing arm curls. His muscles look protean, like putty.

He squints at me. "What do you mean, why?"

"I mean, why? Why bother? Why not just get drunk with

his friends, play pool, go to the hockey rink. Do whatever it is guys do."

"Because he's a wacko, that's why."

"Don't you have criminal profilers? I've seen them on TV."

Michael smiles. He's already spotted his retort. "This is not TV."

Michael has the physique favoured by the town's men; a beefy, overfed quality. Perhaps because no one walks, preferring to jump into their duallies, those trucks with double-barrelled wheels at the back, their hoods flanked with crash bars as if they were going to compete in a demolition derby, and drive down to Chaisson's convenience store rather than walk ten minutes. If you do walk, especially these days, you might find Michael sidling up to you in his patrol car. *Just checking.*

There is a raw power in him. He takes off his shirt in front of Donna, unselfconscious around his sister; he seems not to notice that I am there too. But I have a feeling, although vague, that he seeks an audience.

I ought to be attracted to him. But I don't like blonds, I don't like moustaches. That discounts most of the town's boy-men. His physicality translates into a mental abruptness bordering on the mechanical. He is a ship, a plane. Michael is a person who might do this or that, it doesn't really matter. His thoughts are doors blowing open and shut on gusts of wind.

Still I feel something of his power, the effect he is capable of having, as he stands with his back turned, shirt off, the small hollow which leads to his buttocks visible, a mysterious dark V at the back. Men are given to these cruel displays, I am beginning to understand, but you know they will make you pay to touch them.

The car approaches again. I turn at the sound of my name.

This time I see him. The man is a black nail. Thin, dark-skinned. His skull is familiar. I see my own square forehead.

I run down the street, fling the gate to the house open, and hide in the porch. I watch the car glide past. The driver looks lost, as if he is merely looking for a house containing friends to visit. Long-lost acquaintances.

At the end of May I write my exams — eight of them in a row. I am exhausted by the thought that these separate ordeals will determine the shape of my future.

I am distracted by longings, I don't know for what. I don't seem to indulge in the romantic scenarios of Donna and her friends: me and my boyfriend getting married in a church full of white lilies, me and my husband on vacation in Florida; Disneyworld, Orlando. A nice house on the outskirts of town, two four-wheel drives. A kid I would drive to the same school I myself had attended. Instead my future is a series of tableaux I must somehow avoid: the unmarried young mother; the married young mother; the bank teller; the fish processing plant worker.

In the middle of my exams, the temperature shoots up. I remember Donna's father's prediction — torrid summer. I should be studying for my chemistry exam but I am reading *The Alexandria Quartet*. I found it on a dusty bookshelf. The books ranged there must have been important to my mother because they are well-thumbed, the edges of their pages smudged. I drink them in one after the other: Doris Lessing's *The Golden Notebook*, novels by Leon Uris whose titles I forget, John Fowles' *The Magus* and *The French Lieutenant's Woman*, something by Ursula K. Le Guin.

I neglect my studies because of a more urgent need to

discover who I am and could be, who we all are, through fiction. I don't know this will become a lifelong quest. In Durrell's slim novels, the women are either wafting, hippie-ish, and seductive, or dangerous and seductive. I can't figure out if I would model myself on the free-spirited Clea or the ensnared Justine. In any case these novels are a magical trap. I cannot put them down.

One evening, just as the light is waning, I walk downtown. Cars drift by and I stiffen, but their drivers are merely obeying the speed limit. I will get a frozen yogourt in the air-conditioned mall, I will browse the make-up counters in the chilly Shoppers Drug Mart under the vigilant stare of the women who work there, with their Cruella de Vil eyebrows. I will leaf through *People* magazine then walk back home, my without-purpose gait catching the eye of the man at the wheel of the car I do not notice, the car that drives slightly faster or slower than the others, a sunglasses-wearing figure who, as he draws past, watches me grow smaller in the rear-view mirror.

2. THE ORDINARY YEAR

frazil ice

Fine spicules or plates of ice, suspended in water.

DECEMBER 23RD

"Go! Ice axe break your way down!"

Suzanne and I hurl ourselves down the sheer face of a glacier. The task of the moment is to break our fall with an ice axe. This is a fearsome instrument, usually only seen in horror or climbing accident films. If we get it wrong we will end up in the aircraft hangar at the bottom of the ice slope. We throw ourselves, twirl, and scrape the axe down the ice until we slide to a stop. This is hard on the extremities. I have Monet water lily bruises on my legs and arms.

Winter field training takes three days; we learn how to abseil into a crevasse and get ourselves out, how to rescue someone who has fallen into a crevasse, and how to erect and dismantle camp — putting up pyramid tents, cooking on a Primus stove, using methylated spirits to light it — as well as how to load a sledge and attach it to a skidoo.

Field training is a nightmare of ropes and a metal puzzle of "jingly janglies" — karabiners, jumars, and assorted climbing equipment. I discover I have knot dyslexia — I can barely tie a figure eight. I am the slowest person in our group. The outdoors types stand, one leg flung out, bearing the weight of the jingly janglies on their hips, waiting for me to get it right.

We go out in the field for a couple of days, pitching our pyramid tents up the glacier, called Vals, after those distant pistes Val d'Isère and Val Thorens, and which curves above base over the flank of Adelaide Island. We shovel snow to anchor the tents, we melt snow to make dinner on the Primus. We hunker down for the night in our Rab mountaineering sleeping bags with two hats on and a litre of water to drink.

Inside the tent Suzanne and I share, the world turns tangerine, thanks to the canvas. We have to calculate each inch of our space; there is an interior décor to any pyramid tent as regimented as an IKEA showroom: manfood boxes with food and utensils in between sleeping bags, Primus stove on top, gash in the corner by the opening, sleeping bags on three layers of fleece with the head placed downwind from the wind chill, boots outside in the valance, Tilley lamp suspended from the teepee ceiling.

That night it is cold, but not very cold for Antarctica, minus ten or twelve. I wake with my head against the tent canvas and have to manoeuvre it so that a layer of sleeping bag protects it from the wind. All night this is the only sound. There is no regularity, no familiar voice to it. It's the sound of the katabatic winds, the sound of gravity itself, maybe.

Training over, we descended to base for Christmas Eve. Falling as it does in the middle of the manic summer season, Christmas at Base R was low key. Pilots were still flying in the field (wearing Santa hats), tent parties were holed up a thousand kilometres away with only four mince pies between them to celebrate, and science parties were flying in from the Falklands.

But Christmas dinner was taken seriously. The five-course meal for ninety-odd people used a mixture of frozen and cherished fresh ingredients: prawns, avocados, turkey, stuffing, roast potatoes, Brussels sprouts, cranberry sauce, trifle, cake. For the pièce de résistance the cook baked a giant mince pie on top of which little plastic penguins skated.

The two chefs, a woman in her forties and a young Welshman in his late twenties, were base celebrities, their exploits charted in a haze of snapshots posted up on the kitchen bulletin board —

gala dinners, costume parties, giant birthday cakes in the shape of whales or even, in one, a red Twin Otter. The shelves of the kitchen were lined with celebrity chef classics — Jamie Oliver, Nigella, Moro. The chefs worked long hours, doing prep and menu planning while blasting house music or, occasionally, opera. The previous year's overwinterers treated the kitchen as their personal space. In the winter period only one chef stayed on. On the chef's off-duty days, I was told, the winterers taught themselves to cook a three-course meal for twenty-one people.

The dining room had been decorated fastidiously with Christmas crackers and bunting. A syrupy sun coated the tables as we sat down to dinner, fading the already bleached Union Jacks pinned on its walls. Photographs were dotted around the room — two orcas in the bay, their black heads just breaking the surface, the Dash coming in against a watermelon sunset, lights blazing, a team of huskies pulling a sledge across a diamond ice field from the days of dogs and sledges.

The aircraft aside, these photographs could have been taken in the fifties or the seventies. Apart from Gore-Tex, the detail of Antarctic life had changed little in fifty years it seemed, whether it was the Nido powdered milk, which gave us constipation; the Guinness-in-cans; the dried coriander, dried onion, and chewy frozen salmon steaks; the crumbling frozen cheese; or the dog-eared climbing magazines.

Men in shorts and sandals gathered around the juice dispenser before going to change for dinner. The men on base dressed for the beach, wearing shorts, T-shirts, and sandals — perhaps they had genuinely become so acclimatized to the cold outside that indoor temperatures stifled them. None of the women wore make-up, jewellery, or perfume. We all dressed the same, in BAS-issue fleeces, moleskins, identical wool-mix hiking socks.

This all changed for Christmas dinner. Field assistants filed in wearing smart suits, the women on base wore cocktail dresses they had purchased a year before at River Island or Zara. After

Christmas dinner we played Monopoly and Trivial Pursuit in the bar until three in the morning, in an extended version of the British family post-prandial Christmas Day, minus the television. So much of base life was a simulacrum of the mothership moored so far away in the northern hemisphere: the pub, the dart board, the pool table, the Sunday roasts and fancy dress parties.

By the time I left the dining room, it was two or three in the morning. Downstairs I opened the door to the veranda only to find a solitary skua, looking up at me expectantly. A full-grown skua in flight is a fearsome sight and a possible threat to a human, but there was something of the reckless adolescent about this one. I remembered now: this was a half-tame young bird named Bubba. His photograph was on the base staff bulletin board. He had been known to walk straight through the door, I'd heard. He'd probably join the cafeteria queue if he could find the stairs.

Bubba hopped about. He looked impatient.

"Hello, I didn't expect to see you here," I said, uncertainly. (I had never addressed a skua.)

Bubba flew onto the veranda rail and regarded me at eye level. He turned his gaze to the south. He seemed to be beckoning me to do the same. We both gazed out over the bay, where a shaft of sun struck Jenny Island's fortress flanks with a single sword of silver light. Bubba cocked his head at me, his seed-like eye evaluating me. After a while I said, "Happy Christmas, Bubba," and went to bed.

DECEMBER 28TH

I go for a walk around the point, the small peninsula which shelters base from north-easterly winds. I hear what I think are human voices but which turn out to be birds and the occasional seal. I look at the horizon and struggle to understand its emptiness. Surely there must be something out there, but what? Some fabulous hidden city, an undiscovered Paris of the Antarctic.

Piñero Island towers on the other side of the bay. It's impossible to tell how big these mountains are. There are only four colours: blue, white, gold, graphite. I can feel my retina adjusting to this reduced palette. Already I see more detail in the colours here — streaks of orange in the gold, the citron yellow that fringes it. I miss trees, the variation they provide, their succour. Some people on base haven't seen a tree in two years.

How to stop this constant needling desire to be somewhere else? The delusion there is another place where we would be more content, more ourselves. I am on the defensive here, but what is it I am afraid of? Why don't I allow myself to sink, to dissolve into experience?

I'd forgotten about New Year's. Suzanne confided that she had also lost track of the normal calendar we inhabited in the northern hemisphere. "I don't even know what century I'm in, here," she said. "Everything's still Edwardian."

She was talking about the pyramid tents and sledges, the skis and their tins of antique varnish, the lengths of reindeer gut used to tie the sledges together, the Primus stoves and Tilley lamps. The Antarctic must be the only place in the world where technology and capitalism had not conspired to invent improvements over a century. "We use this gear because it's still the best," one of the four Marks on base (all field assistants) said, when I quizzed him. "The only real advance in polar kit has been in clothing and sleeping bags."

On their march to the South Pole in 1912 the Scott polar expedition wore wool undergarments and reindeer fur mitts. Their boots were finnesko — felt lined with straw. Because they were manhauling and sweating more they deliberately wore less fur than

Amundsen's team. But if they stopped, as in their final nights, held back by bad weather, they became cold quickly. Even if you are wearing a Rab Infinity Endurance or Polar Alpha jacket (the names say it all), you have little chance in exposed conditions, especially overnight in the Antarctic.

Just as Scott's did, our pyramid tents weighed over one hundred kilograms each, as we discovered during winter training. Lashing them to the sledges was an art. The manfood boxes easily weighed in at fifteen kilograms a piece. The Antarctic world is one of physical toil and heft, but this is not the reason women were barred from it for so long. In training Suzanne and I were left to erect the pyramid tents on our own — we might have no help in the field. We lifted and broke camp unaided.

Time had also stopped in social relations, for lack of a better term. We had reverted to an age before mobile phones and Wi-Fi. Our entertainments were Victorian. On base we played board games and did group crosswords and pub quizzes, before going skiing as a group, like children.

"I kind of like it," Suzanne said, as we drank tea in the dining room that night. "It reminds me of when I was a child, before any of this stuff was invented. We used to play charades once a week at home. Now we watch *Downton Abbey*."

Suzanne had taken quickly to life on base. She was an outdoorswoman at heart, reared against the lean winds of Cornwall, a hill walker and rock climber. Crucially she was comfortable in groups and with the rollicking banter that fuels pub talk.

Very quickly groups were established. Suzanne made friends with a couple of the Andys (there were four Marks and six Andys on base) and the younger scientists. I was only a couple of years older than Suzanne, but in the Antarctic bubble the difference was significant. I found myself gravitating to the forty- and fifty-year-olds, the senior scientists and pilots.

Suzanne and I didn't have much time to think, let alone write or make art. Days were occupied by briefings — "sit reps" — fire drills, obligatory group activities such cleaning out the food store, trips into the bay in the RIBs to collect water column samples, talks, and lectures. I took to running on the runway every day, accompanied by a walkie-talkie in case one of the planes was about to land. Several times my walkie-talkie would squawk to life, and I would hear the communications manager's urgency: "Plane, incoming. Bravo Zulu Zulu. Leave runway immediately. Repeat, leave runway . . ." I would look up mid-runway to see a red speck in the sky. By the time I had reached its edge the speck would be on the deck.

The days I lived after stepping into what people on base called the Antarctic bubble would be some of the most vivid of my life. We forget so many days of our lives, days blurred by routine and unawareness. We have to believe there will be more of them; it would be exhausting to live every day as if it were the last one of our lives. Yet in the Antarctic, I was beginning to understand, such intensity was expected, even necessary: most of us had only one shot at the continent. "Time is different here," Mark the field assistant, who had stopped with me on the ski slope on my first night in the Antarctic, told me.

There was no stately progression to time in those months. Time froze or lunged forward, like the outlet glaciers that coursed down the rims of the continent. It could be because we lived in two time zones on base: local time, three hours behind the UK, the same as Stanley or Buenos Aires, and "Zulu Time" (equivalent to Greenwich Mean Time, GMT), which all aircraft operations and communications were conducted in. Perhaps it was down to strange regimes of day and night — stalled sunsets, three-month-long days, three-month-long nights.

I realized I been programmed by a life in the temperate latitudes of the northern hemisphere to expect that the sun would rise in the east and set in the west, and that night would follow day. To be released from these mechanics, which I was unaware of having internalized, was to be freed from a machine I suppose I had tired of. Part of my euphoria was explained by the fact that there was just so much light. It was chemical, but also psychological.

Language was also different in the Antarctic. Those first nights in the bar I learned the truncated idiom of base. There is a British linguistic habit of trying to diminish the power of exotica by shortening names. Even the shortest words are ruthlessly guillotined: leopard seals, the only carnivorous seal in the southern hemisphere and one to be avoided as they can easily kill a person, are *leps*; scientists are *beakers*, named after the *Muppet Show* character; mechanics are *mechs*. Then there are the equally amputated place names — *Monty* for Montevideo, *Punta* for Punta Arenas. The expressions of everyday interaction permitted on base meanwhile were all hyperactive, welded to exclamation marks: "Magic!" or "Super-duper!"

The pilots spoke in a Morse code of their own: "BZZ will depart Site 8 at oh eight hundred Zulu" was a typical phrase. My all-time favourite word was *uplifted*. This was the term used when the Twin Otter finally came to get you in the field, to bring you home to base.

That night the bar filled with more people just arrived on the Dash, wearing the familiar expression of the base newcomer, even if they had been here before: a composite of exhilaration, fatigue and disbelief. Only forty-eight hours ago they'd stood in the departure lounge at RAF Brize Norton in Oxfordshire with its ropey vending machines and tweed-jacketed newspaper kiosk owner, and here they were in the Antarctic, squaring themselves up to nearly eight weeks' work in an underground ice cavern and the punishing winds of Berkner Island.

Everyone talked about the ice coring team long before their arrival. They were the scientific superstars of the season, these forensic archaeologists of the Antarctic. That night there were three: the chief scientist, Xavier, who at fifty was already a legend in ice core drilling; Jonah, a tall engineer; and Gemma, a young glaciologist. I'd seen pictures of Xavier in the halls of Cambridge HQ; in them he looked like an auto-body-shop Santa, with his orange boiler suit, his knit cap, and his ice beard, the skin on his cheeks bloodless and pinched.

"It's always fun to watch their beards grow," Gemma told me, when I asked her how she managed for so long in the field with men — young women are not common in ice core drilling. "We have competitions. Or, *they* have a competition, hopefully. My role is to measure their beards every day. Anything to keep yourself from going mad in the refrigerator."

The ice coring was done from an underground chamber where the drill and the computerized equpiment are housed. Here, Gemma told me, the temperature was between minus ten and minus twenty-five, colder, often, than at the surface. The team would surface to cook and sleep in pyramid tents on the blasted plain of Berkner Island.

Xavier was a popular figure; everyone seemed to want to talk to him. I chose my moment to catch him on his own. He turned his brown eyes on me and there was a note, very faint, of wariness. But he was a generous conversationalist, and for nearly half an hour he explained the background to the Berkner project.

"In ice, depth equals time," Xavier began. "It's all about time, really. Another thing about time and ice cores — in coring, time goes backward, not forward. The further down you get, the further back in time you go."

Antarctic ice cores show that there are eight cycles of warm and cold roughly every 100,000 years. "One cause is a change of external input, such as a variance in the earth's rotation of the sun," he said. "The other is a release of carbon dioxide: in the pre-human world this would have been caused by the retreat of ice sheets and subsequent warming of boreal landscapes, perhaps leading to methane release. Or a failure of the carbon dioxide sinks in the oceans, for whatever reason."

"Then what happened?"

"Earth warmed for five thousand years."

"Why did it then stop warming?"

"We don't know." He put his drink down. "Wait a second and I'll get something."

He returned with a sheet of graph paper. How happy scientists are — really they look beatific — when asked to explain something.

"Look at this." Xavier talked me through the red spindly graph lines on the paper. "What the ice cores also show is that major changes have taken place in the earth's climate within timescales we thought were too short to accommodate such revolutions."

I looked again. The core graph looked like a cardiogram; the tops of the red-spike heartbeats were temperature rises, some of them drastic, followed by a fall into cold.

"Some of these represent a rise of as much as ten degrees in about fifteen years." He pointed to the top of the spike. "It must have been very uncomfortable to be alive, just then."

"When did this happen?"

"Within the last 100,000 years. Humans would have been around to experience the final spikes, albeit in hunter-gatherer communities."

I stared at the spikes, flummoxed by the rapid change they signalled, in contrast to the slow time we associate with "normal" climate change, the slow time of the earth's rhythms, those grandiose processes of glaciation and melt that Max and I spoke about on the ship.

"And in the last four hundred or so years in particular the climate has been remarkably unchanging," he continued. "This is what has allowed us to depend upon planting seasons and yields, to extract the planet's energies, and to grow our population."

His conclusion was obvious. Now we are living on the rising

vertical of one of Xavier's spikes and no one knows where the apex will lie, at which degree the temperature will peak, or if it will ever fall.

After talking with Xavier, I retreated to Lab 7, clambering once again over silver boxes of drilling kit and instruments, to sit at my desk and stare out the window at the glacier.

Many times in the coming months in the Antarctic I would feel a subdued chastening which was not so much despondency as a sobering longing on behalf of the planet. For the first time in my life I was beginning to think of the planet as an organism whose well-being I could effect. I had considered this before, of course, but in the abstract. In the Antarctic I felt closer to the planet than ever before. It was almost as though I could hear its pulse. Was this just a fanciful projection of mere human guilt?

I was learning more in a matter of weeks than I had in entire years. I was relieved to have been released from the narrow concerns of the day-to-day. I was no longer an observer in a global drama but a participant. I considered the paradox of our geographical position. In the Antarctic we were beyond the human, temporarily protected from it. The Antarctic is changing, and will change, but it is out of the equation for experiencing the human effects of climate change, because it has never sustained human life. The disruptions and mass suicides foreseen by climate apocalypticians will never take place there. We were at the epicentre of the crisis, yet simultaneously safe.

Before I'd left the bar, Xavier had passed me on to a physicist to speak to that night. The physicist was a tall, genial man from Durham University. He was studying neutrinos. These sounded vaguely familiar, in the family of quirks and quarks and other subatomic matter I knew little about.

The physicist told me how a star collapses: for tens of millions of years it has fed on the fusion of hydrogen to make helium. But when all the fuel is gone the star collapses in a sliver of a second.

It sends neutrinos into space, soaked in clouds of star-rubble. "Remember, this is happening when man is gnawing the bones of tigers and competing with strange breakaway groups of hominids," he had said. By the time the neutrinos reach earth we are staffing international space stations. They pass through our bodies undetected and come to rest deep underground, in defunct coal mines, in the frozen subterranean lakes of Antarctica. Stars, the physicist said, or their descendants, are moving through us, always.

By December it is too cold to flee his rages and hide in the bat-infested cabin in the woods. Deep snow pockets the spaces between the trees; it will be more difficult to outrun him, panting behind us, rifle in hand. He is fit, still, but the drink and the prescription medicine take their toll. He sits down in a heap and puts his head in his hands, the rifle leaning against a tree beside him.

One night, close to the end, he comes home and hits her. He threatens to get the chainsaw, which he keeps in the bathtub. "I'll tear the place down," he yells.

I go to the kitchen drawer, take out a knife, jump on his back and shove it in his shoulder, hard. Blood spurts over my nightgown.

He doesn't know what has hit him. Then my grandmother is peeling me off him, even though I cling to his back. I almost have the knife lined up with his throat.

"Get! Off!" My grandmother's hand. She pulls me away by the scruff of my neck, like an animal.

Why is she trying to stop me from killing the madman? "The chainsaw," I say. "He'll kill us all."

He collapses that night, more from drink than blood loss. But my grandmother is a practised backwoods medic, and soon she has made a poultice. She puts my nightgown in the sink and burns it. "He won't remember anything tomorrow."

The ice locks in a cordon around our island. For much of our island's history the lake and sea ice spelled death — falling through the winter ice, crossing lakes or the inland sea on foot or by sleigh. They drowned, or were saved, pulled out from the disintegrating ice, only to later die of hypothermia. Sometimes everything would go through: horses, a sleigh, in later years a skidoo. If you could get the person or animal out quickly enough and into warmth, they might survive. Men used to travel roped together, like horizontal mountaineers, so that they could pull each other out walking across the ice that covered the two-kilometre-wide gut of water that separated our island from the bulk of the main island.

Years later, after I had left and was living with my mother in the timber baron town, I returned in winter. The inland sea was open, barren and blue. A film of ice scalloped its edges. "It hasn't frozen this year, or the last," my relatives told me. They shook their heads at the memory of the years when they'd walked on water.

It is minus twenty. We limp through another Christmas. For two weeks I can't get a straight answer out of any of them.

"I'm hungry!"

"Well then make yourself something to eat. There's a good girl."

"Will you help me put on my snowsuit?"

These appeals are met with a drugged silence. They are all sitting, him, her, her sisters, at the kitchen table, playing cards. A quart of rum sits in the middle of the table, half-full. The smoke from their combined cigarettes is not elegant, it does not curl as in films where card sharks play poker games in wheezy dens. I can't breathe in this lung cancer sauna. I struggle into my snowsuit and go outside. One step outside, the cold bites at my cheeks.

I take the dog for company. We walk across lakes, through boggy taiga traced by thin veins of mining roads, flatlands corrugated by drumlins. I am eleven years old. My schooling at the local primary is rudimentary; I do not yet know that the polar regions, the plates of ice that top and tail our planet, exist. Another thing I do not know: soon this chapter of my life is about to come to an abrupt and dramatic end.

I look for my dog, Lindy, in the black cage of trees. She is nowhere to be seen. I fear she will meet the fate of our other dogs, caught in bear traps and frozen to death overnight, dragged off into the woods by an unidentified creature — a lynx, perhaps, or a wolf. The anxiety of potential loss is in direct proportion to the intensity of the thing loved. Even then I had an inkling that life was about this: about testing the limits of loss, how much of it we could take. That this was part of the intentions of an external intelligence. Through loss it would determine who was fit to live and who was weak enough to die.

I call and call for her, but she does not return. The dog is possibly the most intelligent of all of us, the first one to defect from our republic of cold sorrow. But she shows up a day later, the glint of hunger in her eye.

3. EASY LIFE!

shuga

An accumulation of spongy white ice lumps, a few centimetres across, formed from grease ice or slush and sometimes from ice rising to the surface.

DECEMBER 31ST

New Year's Eve. A low mantle of cloud hovers above the horizon. The cloud and land press the sky until a thin blue line sandwiched between two grey-white smudges is all we can see of it.

Niall the met man from Northern Ireland explains that this is a typical Antarctic effect caused by the temperature differential between the cold fission of sea ice and the moisture from summer melt ponds formed within the ice itself. This is the reason cloud hangs much lower in the Antarctic than anywhere else on earth, he says.

Base is busy with preparations for the New Year's Eve party. There will be a formal sit-down dinner, followed by a set by the base band. Its musicians have taught themselves to play their instruments over the previous winter. Melissa the doctor has learned to play drums, which she does very convincingly with the stone-faced concentration of Charlie Watts.

The party is held in the sledge store, where the skis, mountaineering equipment, and clothing are kept. After a while I go outside to see the colours of early morning. Standing next to a rubbish skip is a figure in a parka. We stand in easy silence for a few minutes, listening to the thump of music coming from the sledge store, whose windows had been taped up with bin liners to simulate night.

Our eyes are fixed on the icebergs in North Cove, stunned by the sea ice and the colours of the sky layered with sherbet shades — cold pink, mandarin, lemon. It is three thirty a.m.

"Nice scene, isn't it? Myself, I never tire of the skies down here."

His voice is level, certain. On closer inspection his face rings a bell. I have seen his mug shot next to the tagging board. In the photo he is dressed in a floaty jacket of a material I learn is called Ventile, a favourite among pilots, and made from parachute material, lightweight yet very warm.

The photo on the tagging board is sepia-toned. In his jacket he looks antique also, a hero from another, more legitimate, era.

I had heard of Tom. Several of the scientists had recommended him as a friendly and helpful pilot, if a bit of a chancer, as they all were. He didn't look particularly daredevilish. He was lean, in his early fifties, perhaps, with hair that would have once been dark but had now greyed. From one of his wrists hung a chunky watch with many complicated dials. He saw me looking at it.

"Aviation watch," he held it up to me, so I could inspect at closer range more moving parts I didn't understand. He was just back from a trip down to the Amundsen-Scott base, also known as the South Pole, he told me, and this was why we hadn't met before.

We leant against a skidoo and talked. Tom was Scottish, originally, but he'd spent many years as a bush pilot in the Yukon and the Northwest Territories, and his original buzz had been intermingled with flat Canadian vowels. His lean frame had a reticence about it, also a confidence. His eyes had a hawkish, evaluating glint: he scanned the terrain in front of us as if the ground might rise up and take a swipe at him.

I was fascinated by what these pilots could do — taking off and landing on snow and ice in the most volatile weather conditions on earth, finding a site in the middle of a featureless ice sheet with only the sun and a GPS for guidance. They regularly landed the wheeled Dash 7 on an ice runway, something the pilots apparently called "wrestling a greased alligator with one hand tied behind your back." Tom flew both types of aircraft on base. I asked him which he preferred.

"The Twin Otter, now that's the classic bush plane. It's quite . . . rustic. You'll find out, hopefully, when you get out on a co-pilot."

"I don't know if I will," I said. "I'm not sure what the field operations manager has in mind."

The field operations manager, known to everyone as "the FOM" was the real power on base; he controlled the movements of all the logistics and science personnel. Everyone on base was subject to the FOM's overarching will.

The FOM that season was a lanky Antarctic veteran named Steward, "with a 'd,'" he was quick to tell us, when we first encountered him. One of the field assistants, a young northerner and one of the six Andys on base, gave me an account of what it was like to be subject to the FOM's all-powerful whim.

"You're in your tent and there's a blizzard blowing a hoolie outside and you radio to say, uh, Stew, when are you going to come and get us? And he says, maybe tomorrow, maybe not, maybe next week if such and such happens. Then again, maybe in three weeks. Easy life!" This was the FOM's favourite phrase, something of a campaign slogan for the Antarctic, inherited from the all-male days of the past.

"Then what happens?" I asked.

"Then Stew puts the phone down and you sit in your tent for another ten days."

The lesson was clear; each morning at sit rep it was drilled into us — in the Antarctic plans were scrawls on a whiteboard to be scrubbed out when the next snowstorm hit or the plane encountered a mechanical fault. As if to demonstrate the principle, the eventual day of my co-pilot mission was scheduled and abandoned three times.

It was dawning on me just how big the Antarctic was, and how long it might take to really see the ice cap. On base a week was the minimum for going out "into the field," because even a two-night trip down to the refuelling station and logistics encampment at Ice Blue, the geographic equivalent of a citybreak to Barcelona or Berlin, could become a two-week trip if the weather took a turn or mechanical problems arose (or both).

"The snow is changing. There's a lot more moisture in it now," Tom said at breakfast the following morning. He told me about the time it took him three tries to get off the ground at Base K, a geological base located on the shoulders of Graham Land, where the Antarctic peninsula joined with the West Antarctic Ice Sheet. "The skiway is on a slight incline." He raised his hand, fingers sloping upwards, to demonstrate an angle. "The snow was grippy. The field assistants with me were pretty nervy after three aborted takeoffs."

"What did you do?"

"I just revved her and revved her until we got airborne." His expression was matter of fact, even bored. For him, I would learn, a machine was a challenge. The objective was to control and master it. Like all the pilots, he was a character, an instantly appreciable presence who punctured his conversation with matchless anecdotes and dramatic stories, accumulated in this specialized job where experience counted. Most of the pilots on base had been flying in polar conditions for more than ten years, I would learn. *There was the time we holed up in the fuselage for a week until we only had a tin of beans between us. The time we had to medevac out a Chilean soldier in a whiteout. The time when . . .* For all their anecdotal prowess, they struck me as loners, these pilots. It could be that the weight of their responsibility kept them solitary. They were never alone in the plane — BAS policy was to always have at least one other person with them, as "co-pilot" — but on base they kept to themselves and were only rarely sighted in the bar. They never came to the weekly film night; they never came skiing.

Tom took an interest in my personal mission. He gave me newsletters he'd collected down at the US Amundsen-Scott base at the South Pole. He had been there to fly an electrical engineer in a circuit of successive nowheres, programming the GPS for the non-places where low-power magnetometers, weather instruments, lay covered in feet of snow.

"Those flights were the limit of the Otters' range, sometimes," Tom said. "Ten hours at a stretch without refuelling."

"How did you do that?"

"At times I flew high, higher than I should have, by rights, in order to save on fuel. Lower levels of oxygen means the engine burns fuel more efficiently. The Otters' ceiling is about 12,000 feet. Beyond that, you risk passing out."

"But you didn't."

"No, thank God. But I've had several 'moments' over the years."

"What do you mean?"

"Moments — that's what I call when things could have gone wrong." There was a reluctance in the cast of his mouth. He wasn't accustomed to discussing failure, or even near failure. I pressed him for examples.

"Once I was showing a geologist the strata on a col down near Patriot Hills. There we were, flying into the face of a nunatak — the peak of a mountain buried in snow — and I realized I couldn't get the power I needed to clear it. I think a crosswind on a col took me by surprise. I did it, but it took maximum power. You don't want to have to do that too many times." His face clouded. Like Max, Tom had one of those quick-change masculine faces which seem so readable, but which keep their actual emotions hidden behind a curtain. I saw Tom's eyes better. They were an unusual grey-blue, like winter woollen blankets.

Later that night, Tom and I took a walk together up to the Cross. At two thirty a.m., sunlight beamed from the horizon, tracing the edges of the glacier in gold.

"By next week, the sun will be touching the horizon and the colours will be pink. By early February, the sun will go over the edge." He paused. "There was a girl — a young woman," he corrected himself. "She died here two years ago."

I had heard about the incident before coming to Base R. The young woman was a diver, and had been attacked by a leopard seal. The seal had likely mistaken her for a Weddell seal, its usual prey, and had dragged her down to a depth of sixty metres before letting her go. She was only twenty-eight.

"She died over there —" He pointed to a place in the bay, near the end of the runway.

"So close to base?"

"Yes. I'm told people saw her — that she was in trouble. They saw her disappear under the water. They got in the RIBs, but by the time they found her it was too late." He gave the horizon a flat stare. "I flew her down to Ice Blue; it must have been about six months before she was killed. She was so — so *thrilled*, you know. At first I thought, She's a bit over the top, this girl. She was so enthusiastic, like a child. But then it reminded me of what I'd felt, my first few seasons down here. The wonder of it all."

"You don't love it here anymore?"

"I do. It's just the novelty has worn off, maybe. But I still love it." His voice was resolute. Perhaps he was thinking of the danger inherent in loving a place which had so summarily killed people he knew. "It's the simplicity of life here I like. Deal with the cold, survive, do the job at hand." He gave me a searching look. "Once you've known this life, you never forget it. It holds you, in some way. It's like a horizon you measure the rest of your life by."

We turned our eyes toward the icebergs grounded in North Cove, their flanks now cast in mint. "Yes," Tom said. "You're here at the right time. The sun won't set now till February."

I felt the pull of the horizon and thought, Now the sun will sink. But the sun remained in the sky, stalled just above the lip of the world, hovering over a sea greasy with congealed ice. This was our world, at four in the morning: mountains, ice, water, sky — all lit in lava, tangerine, mauve, and that dark black-purple of dogs' gums, like a bonfire, when the embers have burnt low.

JANUARY 5TH

The met man scrawls the weather, wind speed and direction, and temperatures on the whiteboard each day. Today it is one degree above zero. The mercury has been hovering at freezing or just above for a week now, and melt is underway. There is no smell — not even the ionized smell of melting water. Only that moon smell of gravel.

A biologist (Karin? Karine?) gives an evening lecture. Her PowerPoint reveals the screensaver on her laptop. We catch a glimpse of two blond children immersed in a lake.

Nothing can grow on the ice cap, Karin/Karine tells us, except moss, particularly a type of black moss that has developed a resistance to the UVA and UVB rays they receive over the three-month-long summer with only a thin veil of ozone to protect them. At the top of the peninsula, she says, are fields and fields of strange waving grass, dotted with perfectly intact carcasses of seals who died ten thousand years ago. She shows us pictures of these seals and they are remarkably lifelike with their dun fir, their toothy death mask grins.

To be in a place with so few terrestrial life forms concentrates the mind. I have been taking life for granted — not only my life, the life force given to me so arbitrarily, but any life form. The next time I see a tree I will look at it properly, I will thank it for existing.

Summer was term-time in the Antarctic. "The University at the Bottom of the World," as Xavier said at lunch, as we sawed through another frozen turkey steak served with tinned asparagus. "It's a tradition from the days of Scott."

In the evenings, scientists gave informal seminars. For these lectures we sat at flimsy cafeteria tables in rows, dressed identically in our BAS-issue tangerine fleeces and moleskins, a paroled polar army.

The talk one night was by Gavin, a professor of marine biology. Gavin was considered a "big picture" man and was frequently invited on *Newsnight* and *Channel 4 News* to explain the complex interrelations between the polar regions and climate change.

"Here human beings and science have slightly different objectives. Science seeks to understand, ideally in a way that's disinterested and detached from politics and economics," he tells us. "For human beings it's less about climate change itself, but finding the solutions. It's about the struggle to find new energy sources, fundamentally. And to ration what we already have."

After dinner I ended up sitting next to Gavin at the dining table. It was nine o'clock and the white polar sun streaming through the windows warmed our arms.

Gavin was easy to talk to, with his loping energy, his cackling vaudevillian laugh. I could imagine him in the seminar room: argumentative, challenging, an inspiration to his students.

"So what will you write about?"

"I don't know yet."

"What is your research process?"

"Everything," I said. "Talking to you now, in fact."

We talked about uncertainty and the creative process. I could tell immediately he was one of those scientists who understood the value of an open-ended enquiry, of groping in the dark.

"But all Antarctic stories are the same stories," he said.

That was true, I replied. "But I'm not just looking for a story. Fiction is a form of abstract thought."

He shot me an alert look. "Based on what?"

"Based on what the individual consciousness perceives of the world."

"Really? Is that so relevant? What about communities, nations, families?"

"Even a consensus is just a concert of individual consciousnesses."

"Why write fiction, though? Why not write a non-fiction account of what you learn here?"

"Because fiction frees you from fact. You move into the territory of metaphor, which gives you a density lived life usually evades. Sometimes you have to lie to tell the truth."

Gavin stared harder. A click — not unlike Max's trap door — took place somewhere behind his eyes.

Polar science is intensely competitive but paradoxically collaborative. Intellectually thrusting characters like Gavin abounded — people with real self-confidence, often backed up by the physical courage and sang-froid required to withstand the captivity and extremity of Antarctic field seasons. Gavin routinely dove beneath the ice in the Antarctic among carnivorous leopard seals and orcas. I tried to imagine being in that darkened water, the surface completely iced over, like a sarcophagus.

"What the ice record tells us is this," he said as we finished the traditional cake and tea served that afternoon at smoko, as morning and afternoon tea breaks are called in the Antarctic. "Our climate has been volatile in the past. But now it shows a steady upward temperature trend that is irrefutable. To our knowledge, such a steady unbroken trend has never taken place since the Carboniferous era, and it's directly linked to carbon dioxide. The concentration of CO_2 in the ice cores just goes up and up. The final thing to prove is that the CO_2 is anthropogenic. I think that's been proven inconclusively.

"Now it's about building a consensus, to do something about it," Gavin added. "That requires political will and also economic incentives. Human beings are not very good at paying for the future in the present. You know you're not going to be around to experience either the best or the worst of what we do now, in this time-frame of our individual lives. Naturally people adopt a live-for-the-moment attitude. We're very bad at delayed gratification."

"We're not very good at no gratification either."

"We're selfish and romantic, basically," Gavin said, his burly diver's arms crossed over his chest. "The Antarctic has long been a magnet for romantics, but also their graveyard."

I was not the only one to stay behind to chew over Gavin's talk. With his New Romantics hair, Ben the marine biologist stood out, not just for having brought hair wax to the Antarctic.

Ben was a winterer — in the Antarctic the seasons also described people: *winterer, summerer*. Although it was summer, Ben was still a winterer, because he had overwintered. Summerers were lightweights, fly-by-nights by comparison. Such class politics were never discussed on the ship, but this distinction was clear after only a few days on the terrestrial base. Winterers were the Antarctic elite.

How's it going out there in the real world? Ben told me this was the phrase he used to begin his weekly phone conversations with his mother. He spoke with her sometimes against his better instincts.

"Sometimes I think it's better just to live in the bubble," he said. "It can be torture here in June or July, when you're living in total darkness and your mates email you pictures of barbecues and parks. You know what it's like to miss someone so much your skin hurts? I think they had it much easier in the old days, when you had one hundred words a month."

Ben was referring to the pre-internet communications system, when the base commander sent one hundred words a month as a situation summary over a satellite fax; this, along with a patchy telephone line, was the only form of communiqué until the late 1990s.

"But don't you think this is the real world?" I asked.

"It depends what you mean by reality."

On base people — especially the previous years' winterers — repeatedly referred to the "real world" as being located somewhere else, usually the UK, but on closer questioning it seemed it could be a spectrum of possible locales: places they'd been on holiday, or had lived or worked were also considered "real." How could they live in a sliding scale of realities, with some realities more "real" than others?

Ben shrugged. "Things don't really change here. I can't explain. You've got to live here to understand. It's as if once relationships stick over the winter, then that's it, no matter who comes on base. I want to get back to the real world because there change might happen."

"What do you want to change?"

"Anything!" He looked euphoric and also crushed.

"Are you glad you went through a winter here?"

"I'll never forget it," he said. "But I'll never do it again."

"What did it teach you?"

"To rely on myself. That I could not be swayed. That I could choose not to." What did he mean? I wondered. What could he be swayed toward, swayed from? He didn't elaborate. He leaned forward and dropped his voice to a near whisper. "You know, all through the winter here I had the strangest feeling — I felt as if I'd been put here as a custodian. As a caretaker. I felt I'd been left in charge of a child. If I wasn't careful, it could die."

In was two a.m. I had lost track of time again. I went to leave.

Tom caught my arm as he was coming out of the bar. "Are you turning in?"

"I thought I'd go up to the Cross to watch the sun that never sets."

"Don't suppose you'd like some company?"

Instinct answered for me. "You know, it's so hard to get any time away from people here; I'd just like to be with my own thoughts for a minute."

It was true; since arriving on base the blare of activity had exhausted me. My thoughts fell as if they were slipping through my mind. If I tried to get hold of them they slid through my fingers, like fish. He had been walking in step with me, then, to the boot room, but he stopped in his tracks. "Sure," he said. "I understand."

Standing at the Cross that night on my own I watched the sun hover at the horizon once more, saw its orange flutes, as if a collection of church spires had been buried in the ice, just beyond the visible horizon, and were projecting themselves. Light was refracted upwards through them: rays of indigo, purple, saffron. If I closed my eyes, this all narrowed and started to look like an explosion — too violent, strictly speaking, to be beautiful.

I shivered. It was three thirty in the morning. I'd been at the Cross on my own all this time, not counting a lone insomniac skua who flew sorties nearby, looking at me out of the corner of its eye. If I stayed at the Cross for another hour, I would see the sun lever itself from its hiatus and climb once more into the sky, watch the cold pastel colours loosen and melt as they gave way to the golden orb of day.

My mother hands me an envelope. It is slightly crumpled, as if it has been shoved through the letterbox in haste. There is no address on it, just my name.

By the river. Come at eight o'clock in the evening on June 16.

He wants to meet somewhere open, I understand. Somewhere we can both walk away.

I walk through vacant streets. There is now an unspoken order in town, that women not walk alone. I don't feel particularly vulnerable, I've decided. I don't feel small, or like a victim or a target, even though I understand I am meant to fear being these things. I actually feel infinitely powerful. If I marshal my store of rage, I am sure I can fight anyone off.

I disobey the traffic lights because there is no traffic, while the few other pedestrians wait at the curb. I think how we spend so much of our lives waiting for an external signal, for something to tell us: *Stop. Go.*

I can smell the river four blocks away; the differential between its cool waters and the waning heat of the spring day gives its glaucousness an edge. I walk along the riverside path. It passes under the giant stanchions of the bridge. Above me the traffic is a muffled roar.

Ahead, I see a dark-haired man, hair cut close to his skull. He comes upon me like no other person in my life, he coheres out of the horizon, as if through a conspiracy of the landscape, the sky.

He stands with his back to me, overlooking the river. He looks neither short nor tall. Even from a distance I detect a posture of civility.

He turns around when he hears my steps. I see a version of my face, or rather half of it — my eyes, certainly, my forehead. The lower half is someone else's. I had thought how my eyes were very slightly asymmetrical, the left narrower and longer than the right, was mine and mine alone. But I have been copied. Or I am myself a copy.

He wears jeans, a short leather jacket. He looks neither old nor young. His eyes are dark, like coal.

"I thought you wouldn't come." There is a rough note, a kind of shear, in his voice, which is also present in my voice.

"Why have you been following me?"

"I haven't; I just had to get to talk to you somehow. I didn't want to go to the house. Your mother wouldn't tell me anything."

It is there immediately: a hope, an indulgence. What some might call a parental tone, but I am not accustomed to it and recoil.

Who are you? The question in my mind seems wrong, so I don't ask it.

"Why are you here?"

"To see you. Look —" a note of impatience, the first familiar thing about him, leaks through. "Can we go somewhere to talk?"

"We're talking here."

His mouth twitches.

He is tall — I settle for this, I decide. His eyes are unusually blue — midnight blue, Elin would say. She had an extensive vocabulary for colour, thanks to the myriad objects she sells. He is lithe, not coarse. From the side he has a strong profile. It reminds me of Abraham Lincoln's, on American dimes.

"What have you been doing?"

"When?"

He smiles. "All these years."

I shrug. "School. Life. Living. You look younger than I thought."

"You pictured an old man?"

"Not old — just."

He gives me a long bare look.

"Well you've seen me now. I don't feel anything for you." This is not strictly true. But I want him to think it is.

What I feel is a strange hum, located somewhere underneath my blood and my brain, as if I have tuned into a lost frequency. The familiar-but-not-familiar look and smell of him make me feel like throwing up. At one point I have to put my hand over my mouth to stop this from happening.

The evening has clouded over. The river behind him is a dull platinum. It's the easiest thing to do. I turn around and leave. I can see my action, the potential finality of it, revolving in his head like a diamond, for all his future: *she turned on her heel.*

I don't say goodbye. I think this is the last time I will ever see him.

He discovers me, by accident rather than design, working at Shades of Light. Elin witnesses our interaction.

"Who was that man?" she says, once he has left.

A solemn look falls over her long, serious face. "Oh, I thought you lived with your father."

This time I meet him in the coffee shop, La Vie en Rose. Its walls are painted a hungry red. The café serves cake and coffee to sullen teenagers like me.

He sits across from me, that strange avid look on his face. It is as if he is imbibing me. I can feel myself slipping down his gullet.

I stare at him, looking for the clue to myself. When I close my eyes, an imprint of his face remains on my eyelids, like a photographic negative.

I think, You have nothing to do with me. A word installs itself on my fork, so when I pick up my piece of cake — which I can't stomach anyway — I see it there: *stranger*. The stranger is the person unknown, unbidden, the person who comes and goes with the night.

"Why did you leave her?"

"Is that what she told you?"

"What do you mean?" It has never occurred to me that my mother would not tell the truth.

"I didn't." He says the words as if they contain something sour. "Leave her." His mouth puckers. "It wasn't — working out."

He sits back in his chair. I wait for him to say more.

I am seized by a feeling. Something dizzying, a spiral. But also a sense of lightness. I have been wrong about the circumstance of their parting and this error frees me.

When we part our attitude is diffident. He is not so much in my power, now. We stand at right angles to each other on the sidewalk. He looks as if he has just chanced upon me on the street, a random girl, and offered to buy me a cake. He casts his eyes around the town, looking for something to spark his attention.

I am anxious not to let him think he has scored any kind of victory. On the other side of the street, just out of sight, the plasma of the river cools with the night.

"Well," he says. "See you around." This time he is the one to leave me standing in his wake.

4. ICE BLUE

nilas

A thin elastic crust of ice, easily bending on waves and swell and under pressure, thrusting in a pattern of interlocking fingers.

JANUARY 10TH

Steward slid his thin head around my office door at ten o'clock last night.

"You're off base tomorrow."

The magic words. I will fly with Tom and the chief pilot, Lanier, in the Dash down to the refuelling base. We will deliver fuel and supplies. Then, if all goes to plan, Tom will change places with Derek, another pilot, and Tom and I will travel onward to the Ellsworth Mountains in a Twin Otter to collect geological samples and radar data to send back to Cambridge.

This morning we hold a briefing while drinking tea in the pilot's office in the lee of the hangar. Inside the hangar, the Dash and a Twin Otter gleam underneath the roof lights. Enclosed, they look larger, like reposing dragons.

At eight in the morning, we get the go-ahead from the met man. Tom, Lanier, and I have a mechanic, also named Tom, on the run with us. Tom the mechanic and I will sit in the seats flight attendants normally use, the ones which look backward with shoulder straps. We will face our passengers: sixteen drums of avtur fuel, strapped down where the seats would normally be.

I think of K, my friend who refers to planes as "bombs with wings" — not a reference to terrorism but to the kerosene-filled wings. Now we have a plane full of it. If we crash we won't remember much.

We taxied out onto the runway and took off to the south. The Dash climbed steeply over the bay. When we'd reached cruising altitude,

Lanier flipped open the cockpit door.

"Come sit with us," he said over the intercom. "It's warmer here."

Tom handed control of the plane to Lanier and went to the galley at the rear of the plane to make us all cups of tea. Lanier motioned me to take Tom's seat. I sank into the comfortable pilot's chair with its sheepskin rug. In front of me was a forest of dials; in the cockpit windows two wedges of blue sky.

I asked Lanier about our altitude, course, how the autopilot worked. We talked about trim, stabilizers, the yaw damper which makes a four-prop aircraft easier to manoeuvre.

"You seem to know quite a bit about flying," Lanier said.

"I guess I've always wanted to learn to fly."

"Well, now's your chance."

"What?"

"I'm going to disengage the autopilot, and you're going to fly the plane."

Lanier was smiling. He seemed to be willing me to refuse or demur.

I put my hands on the controls.

"Just keep the plane level and try to keep it to course," Lanier instructed. "Watch the altimeter and the artificial horizon."

As soon as the autopilot was disengaged I could feel the plane in my hands. It was amazingly responsive. Even small adjustments resulted in an immediate reflex from the plane. It felt like a delicate, flexible horse.

I flew for an hour; Lanier's hands were not on the controls, but he never took his eyes off me and I felt completely certain that, were I to make a mistake, he would recover the situation immediately. He told me to hold the plane steady, and maintain our altitude. As for the plane, it had its own desires. After a while I could feel what it wanted to do, I could feel myself anticipating and correcting its intentions, even the flow of air underneath the wings. The sky was absolutely clear and the air smooth. The four propellers chawed gamely through the cold air.

Tom appeared behind me. "Oh, it's you flying."

I didn't dare take my eyes off the sky in front of us. "I guess you'll be wanting your job back?"

"No, carry on. You're doing a great job. You're a qualified Dash 7 pilot now."

Tom took over and I stepped back into the fuselage. I sat down on top of the row of fuel drums. I was completely shocked that I had just flown a plane for an hour, and exhausted from the effort.

I looked at the land we were passing over. Our shadow followed us, a perfect silhouette of our plane, a cigar with whiskers — the propellers. We flew south, into the sun, over a secession of identical landscapes — nunatak, ridge, snowfield, nunatak, ridge, snowfield.

We were passing beyond the peninsula's neck and into the bulk of the continent. Now there was only ice and more ice. All my life I had been travelling through or over landscapes where, sooner or later, some sign of habitation or human interference will appear, even in Canada. But here we were in an iterative world. Ice bred only more ice.

I returned to the cockpit with cups of tea for Tom and Lanier. They had a laminated chart spread out across their knees. Tom pointed to the map. On it, a small black dot amid grids of whiteness marked our destination.

At 74°51' S, 71°34' W in eastern Ellsworth Land, Ice Blue sounded like a place: of blue, and of ice. Officially it was a "logistics facility." The runway was Ice Blue's most spectacular feature, and the reason it existed at all: a groomed blue ice runway over a kilometre in length and fifty metres wide, marked by black triangular flags.

I remained in the jumpseat for the landing. We descended into a ceaseless plain of white. As we lowered into it a thin strip of blue emerged from the blank surface. The strip grew larger and larger in the windscreen. A glint of sun flashed within it as we descended.

The wheels hit the ice runway. For the first few seconds of contact the plane felt as if it had made a normal landing on gravel or asphalt. Tom and Lanier reversed the propeller thrust on the plane. The black triangle of the nunatak loomed larger and larger. Then we simply stopped.

I was awestruck by their skill. The plane had stopped as demure and unflustered as if it had landed at Heathrow. Later Tom explained the procedure for landing on ice. The brakes couldn't be used, for fear of skid, so the pilots had to employ reverse propeller thrust, wherein the direction of the blades is changed to throw the thrust forward.

I yanked open the door and unfurled the plane's short staircase. A different cold pawed its way into the plane: its edges sharp yet thinned by a constant sun.

Stepping out, I shielded my eyes. The light was so searing it seemed to turn into its reverse and become a sultry, bluish night, like *La nuit américaine*, or *Day for Night* as it is called in English, the filmmaking term for shooting during the day and deliberately underexposing the film to make it look as if it had been shot in darkness.

The wind rustled over the ice runway. We had to shuffle to avoid slipping; we were wearing no crampons. I slid toward an orange structure and two snow mounds. A plastic chair sat outside the hut — the weather obs chair. Beside it two figures waited.

Given that we had just landed an aircraft on a plate of ice — even if I had nothing to do with the manoeuvre — I expected a hero's welcome, or at least a visiting dignitary's. But our reception was slightly resentful, as if we had encroached on a private party. Gigantic Jack the mechanical engineer was down for a couple of weeks' refuelling and runway-clearing duty. He wore a T-shirt, shorts, and mukluks. The temperature was minus ten.

I looked back at the Dash, poised on the sheet of transparent

ice, tied down with guy wires to stop a katabatic gust hurling it across the iced runway. Behind it the nunatak peered at us, like some overlord.

The Ice Blue refuelling team followed me back to the aircraft, where Tom and Lanier had positioned the ramp to allow us to roll out the fuel drums. We started to unload the plane, rolling drums that weighed over seventy kilograms each across the ice.

After a few hours of this we heard Steward on the VHF. "Easy life, guys!" He gave Tom and I the go-ahead to overnight in the melon hut, and continue on to the Ellsworths in the arriving Twin Otter the following day.

The midnight sun streamed through the melon hut's portholes. The round windows and smooth geodesic walls made the hut feel like a space capsule. For supper that night there was vodka and chocolate, courtesy of us. We placed the bottle and glasses in a snowdrift outside the melon hut. Inside we sat in a ring on a sheepskin-covered bench, swaddled in sleeping bags. In the middle of the room two Primus stoves and a Tilley lamp hissed and hummed.

Apart from the runway crew, the other Ice Blue inmates that night were Oddvar, a Norwegian glaciologist, and Eric, his long-suffering field assistant, or "GA" as they were called on base. Oddvar and Eric had just spent four weeks together perched on the Rutford Ice Stream, a large moving ice sheet, their GPS showing them very slowly being carried from the Ellsworth Mountains into the Ronne Ice Shelf.

"Eight weeks, and we only saw *one* bird," Eric was shaking his head as I entered the melon hut. "It was a skua, of course."

Oddvar was pale, almost albino, his hair more white than blond. He had light blue eyes that appeared to fracture in their stare, not unlike a piece of ice shelf when it breaks from the continent.

He fixed me with this flinty look. "And who are you." It was more statement than question.

"I'm the writer."

We had a staring contest for what seemed like a long time. The three field assistants sitting round the table darted their eyes back and forth, as if watching a tennis match.

"It is a difficult place to be a woman, Base R," Oddvar said, finally.

"It's been fine so far."

"Don't you get any hassle?"

"Disappointingly little, actually."

This provoked an especially intense stare. He turned to Eric. "As I was saying, female GAs are so unfeminine. I like women with a little" — here he made a bonbon gesture, like sucking on a lollipop or sweet, finished off with a little kiss. He grinned. Oddvar had one of those square smiles: his mouth opened up and down but did not go sideways. I saw he had small, wolfish teeth.

Eric rolled his eyes. He turned to us. "He has two topics of conversation: vodka and women. No, I forgot — Amundsen."

Tom laughed. "Not him again."

"Why is the Rutford Ice Stream speeding up?"

Oddvar flicked his eyes toward me. Eric looked away. I understood belatedly that work questions were not the thing to ask at that moment. I lost the attention of the others. I could hardly blame them; they'd just broken out of ice jail and the last thing they wanted to talk about was work.

But like all the glaciologists I met, Oddvar was gripped by his profession, by the remote, metaphysical aspects of glaciology's collision between the numinous and the numbers, the pure math of it, so he humoured me.

"That's what I'm trying to find out. I take the data and give it to my PhD students, who are modelling ice sheet behaviour. They'll feed it into the models and, depending on the parameters and the forcings, we'll refine the equations that govern the flow velocity and its increase."

"But is it the warming sea that's causing it to flow faster?"

"That's part of it, but not the whole story. The flow rate is governed by what happens on the bottom of the ice sheet. The base is where the increased heat speeds up melting, which speeds lubrication, which speeds outflow."

Max had told me that much of the behaviour of ice sheets was still not well understood. Looking at Oddvar's maps of the Rutford Ice Stream it was difficult to imagine that such a broad tongue of solid ice was buoyed by a greasy layer of slush, heated to the melting point by the thermal energy of friction as two kilometre-thick ice slid over rock, far beneath us.

"The ice sheets are in constant motion," Oddvar went on. "Just as the continents are, just as the earth is spinning. I always try to remind myself that I'm not standing on solid ground."

I left the melon hut to go to the bathroom. At Ice Blue a pyramid toilet tent sat above a pit of frozen shit and piss called the Craporium. There was no smell, but still it was not recommended to dwell too long on the stalagmites and stalactites that coated the ice chamber beneath me. The Antarctic made you acutely aware of bodily functions. I wondered what I was going to do the next day in the Ellsworths, in a remote field camp with four men and Tom.

There were no snowdrifts or rocks or trees to hide behind there, there was no Craporium.

I returned to the melon hut to a raging argument. "Scott was gay!" Oddvar was shouting. "I mean, I have nothing against gay people . . ."

"Amundsen was an alcoholic," Eric rejoindered.

"Scott was gay *and* an alcoholic!" shouted Oddvar.

This was known as the Romanticism vs. Practicality debate, an Antarctic conversational trope so common yet gripping it was impossible not to join in. It goes something like this: practical Amundsen, mechanically shooting and devouring his faithful dogs on his Formula One dash to the Pole, versus Scott the humane animal-lover, refusing to put animals through such cruelty (despite unwittingly slaughtering a brace of Siberian ponies he imported to the continent, unaware that they would be no match for its supernatural cold), and who insisted that he and his men manhaul to the Pole.

"That race was won on calories," I said, provoking a hum of approval.

"Why must the English glorify those who have failed?" Oddvar shouted back. "Why are they insisting on this bloody romanticism? The only mistake Amundsen ever made was not to die!"

I bedded down in a lightweight tent pitched beside the melon hut that night, an eye mask clamped over my eyes against the light. It was a long time before I could get to sleep. I was safe, among other people, and it was summer. This was not the lethal Antarctic which Scott and his brethren had encountered. Still, as I listened to the hiss of the wind I found myself thinking of the dire final days of his expedition.

I had read Scott's diaries; they were salutary and almost required reading for anyone who sojourned in Antarctica. It is still perhaps the best-known story about the Antarctic: how the British Antarctic Expedition from 1910–1913, led by Captain Robert Falcon Scott, competed with the Norwegian explorer Roald Amundsen to be the first human beings to reach the geographic South Pole. How Amundsen beat them to the Pole by five weeks. Scott, accompanied by four men, Lawrence Oates, Edward Wilson, Henry Bowers, and Edgar Evans, did manage to reach the pole, but died of starvation and exposure on the return journey to their base camp at Cape Evans on Ross Island.

Scott's diary suggests the leader may have lived for a day or so after the two other remaining men alive, Wilson and Bowers, succumbed to death, frozen in their sleeping bags. By then he may not have been able to stand, let alone walk. He may just have lain there, his feet ruined, their soles separated from the flesh and bone.

As is well known, the tragedy of Scott's demise is partly because it could so easily have been different. Only eighteen kilometres separated Scott and his men from survival — the distance from the frozen tent, found a year later, to the cache of food and fuel at One Ton Depot. Had he placed the depot at a shorter distance, or managed to press on to reach it, at least three of the five men may have lived.

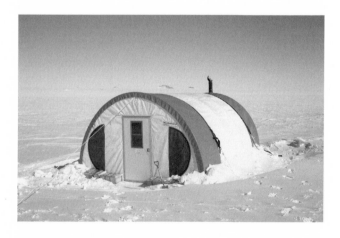

That night in the tent at Ice Blue was the only time I have slept alone in the Antarctic. It was only minus fifteen outside, and it was light. Next to me was the melon hut where four men bedded down, and beyond it, a pyramid tent where the two Ice Blue petrol station attendants on long-term secondment had retreated in favour of the guests. I was hardly alone. Still I stared at the pale canvas wings of the tent flapping in the wind, at the dark slit of the tent's zipper, closed against the outside air.

I wondered at the closing of a life in such a place. How it would feel to know rescue is impossible, that one's body, even, may vanish, eaten by the ice. How it would — how it will, eventually — feel to know you are about to enter into an order of truth that is beyond experience. To be so far from humans, from human history, to die that way, would have seemed to me a distinct death. I had the idea that somehow the abandonment of the self by the self that death entails might have felt less bitter, here. That Scott might have been overtaken by a feeling of protectiveness, of responsibility — that he needs to stay close to his death, or it might leave him behind. As if it is he, and not Death, who is in charge of his dying.

The next morning, Tom and I packed up to go. The Otter had only a rudimentary navigational system, so we would fly using a map marked with GPS coordinates to the deep field camp pitched in the shadow of the Ellsworth Mountains, where two geologists were collecting rock samples.

We took off and headed northeast. After two hours of flying, black crystals appeared on the Otter's windscreen. As we got closer, the crystals became the triangles of mountains. The Ellsworth mountains are a ruche in the otherwise flat fabric of the ice sheet. Located south of the frozen perimeter of the Ronne Ice Shelf, the Ellsworths are part of the West Antarctic Ice Sheet that forms the neck and tail of the Antarctic continent.

We spotted the camp, located in the lee of the Vinson Massif, the highest of the Ellsworths, three black specks on a glaring sheet of ice. We banked and landed in a flurry of snow crystals and suddenly I was hugging two orange boiler suits I had never met before. The boiler suits loaded their samples into the plane while Tom took me for a walk. Bamboo sticks with black flags marked an area safe from crevassing.

The Ellsworths towered above us, black tors against a lordly blue sky. I could see the crevasses huddled in their skirts. Apart from geologists, only rich climbing adventurers came, I was told — heart surgeons and venture capitalists brought by specialized tour companies to pit themselves against this still largely unexplored range.

I lost track of time: we had arrived at eleven a.m.; now it seemed close to midnight. Or was it noon? Was it possible we had only been here an hour? I asked Tom, who gave me a puzzled look. "We've been here six hours, probably more. Don't you wear a watch?"

I was hefting a box of samples into the fuselage when Tom guided my eye back toward the tents. "Look," Tom pointed into the sky. Above us, the sky was black. Two silver haloes ringed the tops of the tents, and the sun. They were not perfect circles, but incomplete parabola, like rainbows.

"Parhelia," Tom said. "Sun dogs."

We both looked at the ground beneath our feet. Snow crystals, light as coconut, flurried over our boots. On base we had the sea next to us, a recognizable dimension. The ice cap on the other hand felt exactly like its name — like living on a hubcap.

We finished loading and clambered back into the cockpit. The snow was packed hard, unlike the slush at Base K: we were airborne quickly, buoyed on the anoxic air of the plateau. I caught a glimpse of Tom's watch; it was two in the morning. The sun gave us no clue as to the time; it merely patrolled the sky in an ellipsis.

We circled the camp, banking at an abrupt angle. My shoulder was forced against the cockpit window. The geologists stood between the two tents, waving furiously. Even at an altitude of only three hundred metres the whole speck of human endeavour had virtually disappeared: the tents, radio transmitter, skidoos, and sledges were only a cluster of particles against the ice sheet.

"It's beautiful, isn't it?" I couldn't see Tom's eyes behind his sunglasses, but I could hear it in his voice, the tug of consensus.

I nodded, although I was not sure what to call it.

"Time to earn your keep," Tom said.

"What?"

"You're going to fly the plane. I've got some paperwork to do."

I put my hands on the controls. The Otter was less difficult to fly than the Dash. It is a rustic, two-engine plane, ideal for polar environments and the bush, one of the strongest and most reliable planes in the world. I fixed my eyes on the artificial horizon and the altimeter, but from time to time I managed to look out the window at the landscape we were passing over.

Below us waves of sastrugi were caught by the sun, turned into shimmering ice snakes. Glaciers hunched, hummocking toward the sea. Over my shoulder, to the northwest, I could just glimpse open water, two narrow channels of blue — that deep, gem-like blue of water in the Antarctic. But this was a mirage. We were too far from open water. On the ice field I watched our shadow chase us. On its edges was a prism of light like a moving rainbow.

We flew for hours. Beneath us, flat white turned into land that looked like a crumpled piece of paper. This was the peninsula and its familiar mountains.

Our flight was smooth. I was getting better at holding altitude. I relaxed enough to look over my shoulder, back at where we had come from. I imagined the perimeter of the continent, sea ice hunched against its shores.

Tom poked my shoulder. I thought he was rebuking me for

taking my eyes off the sky in front of me. I pulled my gaze back. His voice came over the intercom.

"Never look back," he said. "It's bad luck."

He shook his head. I couldn't see his eyes behind his sunglasses. Instead I saw a mirror version, compressed, of me, my hands on the controls, my earphones and microphone.

"But why?"

"Pilot's superstition. Keep your eyes focused ahead of you. You never know what will happen."

We arrived back at Base R at seven in the morning. We were exhausted. Tom powered the plane down and we sat in silence for a moment. The sky was spliced into layers of orange, blue, gold.

"Well," Tom said. "Home for tea and medals."

At home Mark is watching Wimbledon. Mark did his PhD in England; he'd lived there for four years. He spoke about punk, strikes, a whole nation watching itself fall apart. He talked about those years as often as possible, inserting them into conversations. A light came on inside him. As if only there had he been truly alive.

I have never been to England. From television, PBS serials, and the music it exports, I have cobbled together an unseemly puzzle of contradictory images: a grittier place, poorer and conflicted, yet it supports dowager monarchs and disintegrating mansions. I don't know it, but in four years' time I will go to live there, initially for two years, then forever.

My mother comes in the room and stares at the rectangle of green on the television, the two cavorting figures in white. "Those women shouldn't walk alone at night."

"Men walk alone at night," I say. "Why shouldn't women?"

"Men get attacked too."

"Not the same way women do."

"They do," my mother says. "Believe you me. You just don't hear about it."

My mother gives me an opaque look. I have the impression she is trying to muster concern but can't quite arrive at the pitch and temper required. If something were to happen to me, I have the feeling that my mother would not be glad, exactly, but that it would confirm some essential supposition about the universe and its workings for her. There would be a symmetry, a vindication.

We meet again, this time at the Executive, where I work. I can't be seen to drink at work so I decline his offer of a beer.

"What do you do here?"

"Bus tables. Serve alcohol, although not legally of course."

"But you work at that store."

"I have three jobs, technically." I don't tell him about the stables where I work, the horse I keep there and which I have managed to support for several years now.

"How do you keep it all going?" He squints at me. "With school and all that."

"I don't know," I say. I never think about it. All I know is nobody's going to pay for my university. Since my mother married Mark all her money goes to their children, and because the authorities take his salary into account, I don't qualify for a provincial loan.

He gives me a look I've never seen before. Admonitory, but with a shallow note of kindness buried in it.

"Aren't you going to ask me how old I am, what my plans for the future are?" I am trying to be sharp to give some substance to the boggy feeling inside me.

"Not if you don't want me to. Which it sounds like you don't."

"So what have you been doing?" I don't add *all these years*.

"Well, for five years or so I worked up in the Northwest Territories as a bush pilot. Then I flew out west, in British Columbia. For Air Canada."

That might explain the expensive watch. I know nothing about watches but his has a foreign glint — a soft brushed metal I have never seen before but not unlike the pewter earrings sold in Shades of Light.

"Where were you based up north?"

"Resolute Bay. I flew in Greenland for a summer, when Air Greenland were short of pilots. That was interesting — it's covered by a permanent ice cap."

"I know. That's what sunk the *Titanic*, an iceberg from Greenland."

He smiles. There is a hesitation and a delicacy in it I have never seen in any other man's smile. A note of disbelief.

He sits back. His coffee is nearly untouched.

"Do you have any children?" I mean to say, any *other* children. But something stops me.

"Yes, a boy and a girl. They live with their mother out west."

"Why aren't you there?"

"I wanted to come home. I have to go back soon. I can't stay."

I find myself staring at his mouth. A hungry mouth, a shade crueller than my own.

He starts to speak. It is a long while before he stops. I am unused to adults coming out with such soliloquies and listen, blinking.

He tells me about Africa, the Namib Desert and a strange German town where he was based while flying there, about the desert elephant who roam the sands searching for water, the plants that never see rain but which feed on the mist of the cold ocean. "The colours were incredible. How the sun would hit the dunes in certain lights. They looked like they were on fire. As far as you could see, stretching into Angola. A coast on fire."

My mind leaps to life, as if I had been anaesthetized. I can see the desert that locks into the ocean's edge. This landscape I find myself in — the woods gnarled with tightly packed pine, the cryogenic slumber of its endless winter, fails to move me.

"The ocean is cold there. There's a current from Antarctica. When it hits the hot shores of Namibia, it throws up a thick fog."

I am lost for a while, thinking of a place where hot and cold collide in two rigid streams. *Antarctica*, *Namibia*. Their names have the ethereal sound of distant kingdoms, haunts of dragons and ogres.

How differently he speaks, I think — fulsome, descriptive, like characters in novels. My mother speaks in cryptic remnants of intent, in orders and instructions. For the first time I take the estimation of this man I have in front of me as someone separate from me, released from the burden of the word. *Father.*

We take a walk by the river. When he moves it is crisp, urgent. Maybe all pilots move like this, I think. I like to look at his profile hard against the sky and the razor tops of trees. I decide what I like about him: he is certain, he doesn't hesitate. He doesn't act guilty or furtive.

"I thought you were a stalker, or even this guy, the guy who has been," I stop, unable to say it. "When you followed me in the car."

"I didn't want you to think that."

"Do you remember me when I was a baby?"

"I only saw you twice, just after you were born."

"What did I look like?"

"Like a baby."

I laugh.

"Let me show you where I'm staying." He had told me he'd rented the apartment downstairs from a friend. The house was a short walk from the river, one of those prosperous bungalows with a separate apartment attached.

To his friend, a man his age named Mr. MacKenzie, he introduces me. He does not say, *my daughter.* He leads me into an anonymous living room. It contains a rocking chair, a sofa in a checked pattern, an outmoded television.

"How long are you staying?"

"A week, maybe."

Maybe what? Another thing I don't voice. Meeting him is

refining the dual current I am learning to cultivate in my mind, and which seems to be part of being an adult, that of the things you think versus the things you say.

He gives me a look of disinterested appraisal. "You know, you don't strike me like an eighteen-year-old girl. Not that I know so many of those."

"I'm not eighteen yet, I'm only seventeen."

He nods. "Ah. Sorry about that."

"Don't you remember when my birthday is?"

"I do," he nods, but I am not convinced.

"It's a strange age. I don't feel young."

"Do you feel old?"

"Yes, I do."

He is very close to me. Only the table separates us. I think, this is the closest I have been to my progenitor since I was conceived.

I see that he is nerves and surfaces, as I am. Skin. It seems an inadequate sheath for muscles, bones. I want to tear his off, to plunge my hands inside, into the tissue I learned in human geography is called *adipose*. Fatty, greasy, and yellow, flooded with blood. I want to travel his network of veins, like the thin roads that snake through the woodlots of the province; once inside him I will rummage for the code. He has it inside him. I am not sure if it is his, or mine for the taking.

He is not going to let me walk home but I insist. It is still light; the sun does not go down until eight p.m. now, on the cusp of summer.

To the west the sky is gold. I walk past the shabby downtown outlet mall, the Canadian Tire store, into the elm-shrouded streets where dark-eyed clapboard mansions clump like mushrooms in their shade.

5. UPLIFTED

ice rind

A brittle shiny crust of ice formed on a calm surface by direct freezing, or from grease ice, usually in water of low salinity. Easily broken by wind or swell, commonly breaking in rectangular pieces.

JANUARY 17TH

The height of the Antarctic summer. Base is disintegrating. We hear more sounds now — the sound of water trickling in rivulets, full-on funnels, waves and waves of water washing down from base into the bay.

Around the point there is so much life. The elephant seals thump and sigh as they turn their bodies over. I have never seen so much blubber. It ripples. Their skin is tobacco coloured and peels off in giant flakes.

Skua chicks have appeared in nests at the north end of the runway. Their mothers strafe my head as I run. Fur seals are expected soon. The population of Adélie penguins has tripled daily since late December, when we began an informal census. Every day now we see a couple of Adélies tottering toward us in their dinner jackets, like a special delegation of miniature diplomats.

The Dash flies to Stanley and returns laden with scientists. I love the sound of the propellers, the chop-buzz-whir as the planes hurl themselves down the runway.

This morning I was given a temporary new office-mate in Lab 7, Helen, an atmospheric chemist. Thirties, I would say. She tells me she is married and has two children at home who are currently being looked after by their father and her sister. She has short strawberry-blonde hair and dark green eyes. She looks so much more welded to the world than I am.

I take a shine to Helen. Her desk is littered with high-end sunblock and Clarins super-strength moisturizer. She says, "The sun is a killer here, really, don't even put your face out there for five minutes without Factor

50. Have you been told to put it under your chin yet?" I have. Last year a newcomer contracted a third degree burn after being out on the snowfield for six hours with her chin unprotected.

Helen is keen to get out into the field, get the job done, and get home, she tells me. Her husband is less than happy about her five-week sojourn in the Antarctic. "He seems to think I'll be seduced by a handsome pilot and never come home."

"Has that ever happened?"

"I suppose one did try, about ten years ago."

"What happened?"

"I married him." He shows me photographs of her pilot husband, an ex–Antarctic bush pilot who now hauls 737s for British Airways.

Helen's arrival reminded me that I hadn't spent much time with women on base, apart from Suzanne, who had just left base to return to the UK. Not for lack of them — women made up about thirty percent of base population that summer. Perhaps it was demographics rather than gender that kept me from making friends with other women; the scientists were all very young. Even Melissa the doctor was five or six years younger than me.

The history of women on base was relatively short. At Conference I'd learned that BAS' programme in the Antarctic was off-limits to women until the 1990s, a proscription that was on a basic level about pregnancy — the medical dangers it posed, theoretically, and the difficulty of extracting a pregnant woman from the Antarctic in the winter. (Although a number of live births have taken place on the Argentine and Chilean military bases at the tip of the Antarctic peninsula.)

Women's introduction to the British Antarctic Programme was gradual. In 1986 at Signy, the sub-Antarctic island base, the first female scientists spent the summer; in 1993 the first woman spent the winter, and in 1997 the first woman overwintered at Base

R. Since then there had been a gradual increase of the number of women on British bases.

"There's no longer any novelty, really," Melissa the doctor said. "This year all of the doctors are women. We've had a woman plumber, even a woman pilot. It's not all about sex, you know." She gave me a dark look, as if I might be a proponent of this delusion. "In winter, you've got this overpowering sense of responsibility for your well-being, for everyone's as well as your own. You function as a unit."

"A family unit?"

"More like siblings. That's how I think of the guys I wintered with. They're my brothers."

I remembered advice — warnings, really — I had been given at Conference. It's no place for dark horses. People like you to be real on base, to be readable. Honest, straightforward. The message was that as a woman you needed to be good company, fun, cheerful. Seductresses are bad news for Antarctic communities, dividing men, and women too. Here we were a benign army, only of any use as a unit.

We were sitting in Melissa's office. It was like any other GP's office I had been to, with Post-it notes on the bulletin board, a small dispensary behind locked glass doors, and pharmaceutical manuals ranged along the wall. But a pair of crampons occupied a corner of the room, next to a ski mask and a pair of polar-issue down-filled gloves.

Melissa exuded gravitas. Doctors necessarily played a dual role — on the one hand they were a member of the team, skiing and drinking with the rest, but they were also party to people's deepest bodily or psychological secrets, and at any point they might be called upon to save a life, often working completely on their own.

Talking with Melissa, I began to imagine a scenario: of being in love with someone on base, overwintering together, but for something to go wrong. I tried to imagine what it would be like to pass seven months voluntarily stranded with someone you are in love with. There would be the evanescence of the place, the uncanny light displays and celestial trompe l'oeils of a polar winter. But what if unrequited love, or jealousy, or betrayal entered the picture? No one could walk away.

I mooted this scenario to Melissa. She looked grave for a moment.

"Yes, that would be the nightmare. Everyone knows everything about you here. You can't even sit at dinner together for three nights running without generating rumours. That's why everyone is so cagey here, at least until the summer is on the horizon.

"The thing I was most afraid of here was sparking some sort of —" She gave a sharp intake of breath. "Passion, which would lead to jealousy. As a woman among men that's always a possibility. Over winter that would be bad news."

I lingered on the word, passion. Its true meaning is located in the Latin verb *passere*, to suffer. Even in its white-hot, requited version, passion is still a suffering, because it can never be truly sated.

FEBRUARY 6TH

The light has changed. When did this happen? One day it is full blitzkrieg sun, the next there is a note of hesitation in the light.

The summer science season is winding down; the ice coring and radaring teams are being brought in from the field; no new teams are going out. The population of base stabilizes to sixty or seventy faces. The frantic pace eases. We linger over meals again.

At lunch I talk to Russell, a marine biologist. He is tall and lanky, kept thin and fit by marathon running on the airstrip.

"Why do you think people on base dislike the idea of the writers and the artists who come here?"

"It's not only the writers and artists," Russell says. "It's anyone who isn't a scientist. If you're a scientist, you've fought so hard to get here. Every person here is someone else left off the boat. The Antarctic is the only place on the planet reserved for science, and resources are finite. But also —" He pauses, as if uncertain to divulge a sudden thought. "I think it's because here, no one is anything special. As soon as you think you might be — the arrogant types, the VIPs — you earn the ire of the crowd. Maybe there's a feeling that writers and artists make too much of things that don't stand up to such scrutiny. Plus, there's jealousy of course."

"About what?"

"Of all the airtime culture gets. Meanwhile scientists are working away behind the scenes, being thorough, not asking for any particular recognition, only for understanding."

"Actually I think we're in an age where science has the purchase on truth," I say.

"It might have the purchase on truth but it doesn't have the money or the glamour."

There is a truth here: I've learned that research science is not very well paid, until you are on a professorial salary. Then it occurs to me that people on base might think that because I am a writer I have money, or will make money out of this experience — a comical thought.

Behind Russell's shoulder an iceberg the size of Heathrow Airport sits framed in the dining room window. It blew in on a southerly wind this morning and parked itself at the end of the runway. This is a problem for the planes: the wind is from the north today and the pilots have to use the southern approach. But a mountain of ice stands in their way to land.

"The pilots will just land with the wind behind them, if the chief pilot thinks it's safe," Russell says. "Otherwise they'll wait it out in the field."

"When will the berg move?"

"Who knows? Could be tonight, on the change of tide. Could be next week. Or never. It's the charm of the Antarctic. Nobody knows what is going to happen from one minute to the next."

That night Xavier appeared in the doorway to my office with two books in his hands.

He had returned at some point in the previous day from nearly a month on Berkner Island. He looked just like the pictures I'd seen of him in Cambridge HQ: his rosewood complexion had turned ashen. The shadow of a recently shaven beard stalked his chin.

"I thought you might find these interesting reading." He held up their covers. *The History of Glaciology* and *The Spiritual History of Ice*.

I thanked him, adding, "I've been trying to read the *Annals of Glaciology* but they're full of papers called —" here I picked up a couple of journals from my desk and read the contents page, "'Correlation of Marine Isotope Stage 4 Crytophera Horizons Between the NGRIP and GRIP Ice Cores' and 'Glacier Mass Balance Modelling of the Tibetan Plateau — Mesh Dependence Issues.'"

He laughed. "You'll find these books a little more accessible I think. The first chapter is excellent. It describes when dinosaurs fed on tropical forests here."

Until 180 million years ago, the Antarctic had been part of the supercontinent Gondwanaland. Then, it had been tropical, carpeted by forests of clubmoss trees, the giant ferns of the Carboniferous era. The Antarctic is geologically similar to South Africa; you could see this in its fossil record, Xavier said, in the mineralogy: moist, carbon-rich soil, signifying coal, diamonds, perhaps gold, all buried underneath one or two kilometres of ice.

"One hundred and fifty million years from now," he went on, "Europe and Africa will heave together, their stone edges crumpling before finally welding together in a single continent, consuming the Mediterranean Sea. North America, Australasia and India will collide to form a new supercontinent, while South America will drown in the mid-Atlantic trench."

"And Antarctica?" I asked. "What will happen to it?"

"It will stay where it is. It will become more and more isolated as the other continents move away."

"Then it will have what it wants," I said. "To be alone."

Xavier went back to his computer. Outside the window the glacier gleamed. Two skidoos threaded their way up its flanks, carrying midnight skiers. I settled in my office, my feet propped up, soaking up the radiator's heat, and began to read the books Xavier had lent me.

Reading *The Spiritual History of Ice*, American academic Eric Wilson's luminous study of the Romantics' relationship with ice, reminded me that Shelley, in much of his poetry, was suspicious of ice. It was cold, aloof, haughty, deadly, Wilson writes. Glaciers were generally seen as evil. But with the coming of the Enlightenment the dark spell lightened. Glaciers became, in Wilson's words, "repositories of magic." Science and art combined to describe and capture the mystery of the allure of ice, and its changing fortunes.

Goethe, meanwhile, believed that science and art were a mirror to each other, Wilson notes. "Like art, nature is a lawful, outward pattern of inner, unruly energy; it is disciplined and extravagant; it is purposeful and purposeless. . . . While the groaning ice of the Middle Ages and Renaissance was possessed by devils, the sublime ice of the eighteenth century was haunted by spirits connecting human souls with cosmic powers." Glaciers, Wilson writes, were "ambiguous immensities of rectitude and weirdness, necessity and violation," then were transformed into spirits called *daimons*, "familiar spirits connecting poets to life." These translator spirits guided poets in their quest to know and represent nature. And so, Wilson concludes, the modern sublime was born.

It was three in the morning when I stopped reading. In my window the black peaks of Reptile Ridge — so called because its spiny back looks like that of an iguana — loomed, tipped with the fire-glow of the perpetually setting sun. There was no sign of the midnight skiers but they could have returned long ago, driving the skidoos in the blue twilight.

I walked through the silent corridors of Bransfield House. In each of the laboratories silver equipment boxes were stacked, just returned from the field. Inside were ice core driller bits, radaring equipment, echo sounders — the instruments of the search for the frozen lakes trapped kilometres beneath the surface of the ice, the accelerating ice streams, the fracturing glacier at Pine Island.

All base was silent except for the hum of the generator. Only Bubba the skua — who I knew would be waiting for me on the other side of the door — and I were awake. I liked base best at this hour; deserted, it felt fully extraterrestrial, a landlocked spaceship held fast in the ice.

I swung the door open to say goodnight to Bubba and recoiled in shock. I put my hand out into the night air and felt a substance thick and dark, like the pelt of an animal.

It was night, or a version of it. The skies were changing quickly; gone were the days of blaring sun and the lava fields of the stalled sunsets we witnessed in early January. Out of a cream sky a half sun showed itself. It hesitated above the horizon, as if undetermined whether to rise. Its light was glaucous and stringy, like albumen.

In two weeks' time the planes would begin to leave; first the Otters, then the Dash, after flying one last sortie to the Falkland Islands. Then the base population would be reduced to those of us awaiting the last ship of the season, the *Ernest Shackleton*, due in late March.

I reached for my little plastic name tag and moved it from the *On Base* section to *Pit Rooms*. I said my customary goodnight to Bubba and went to bed.

FEBRUARY 7TH

Today was Tom's turn to lecture at the University of the Bottom of the World. He talked about his recent trip to the South Pole, illustrated with photos.

He tells us there are actually four South Poles: the Geographic South Pole at the southern end of the earth's rotational axis — this was what Roald Amundsen first reached in 1911. The Magnetic South Pole is where the lines of force of the earth's magnetic field converge. This moves, or "wanders," every year and is currently somewhere in the Southern Ocean. The Geomagnetic South Pole is a theoretical point that marks the southern end of the earth's geomagnetic field. It also moves, and is currently somewhere near the Russian station, Vostok, in southeast Antarctica. And finally, there is my favourite, the Pole of Relative Inaccessibility — the place farthest away, in all directions, from the Antarctic coast.

We get to see the American Antarctic. Tom witnessed the American summer traverse from the McMurdo base on the coast to Amundsen-Scott at the South Pole. He cycles through photos of lunar, lumbering vehicles, hybrid tractor-combine harvesters, snow caked on their treads, hauling shipping containers on giant sledges.

"It's a little different down there," he says. The American Antarctic is slick and big, predictably: a far more corporate approach than our modest Edwardian venture. There is an ATM linked up with the mainland banking system via satellite. Their accommodation looks like a large university block building in, say, Wisconsin. They have a bowling alley and basketball court.

We stare in wonder at the potbellied C-130 landing on the ice runway, the tractor-combine harvester snowblowers like winged steel dragons, the Moon Unit three-storey accommodation blocks, their giant probing telescope, their lavish ice runway, the Coors and Miller shipped by crate all the way from California while we make do with Uruguayan beer procured in Stanley. We can't help but look around at the walls that enclose us: the Portakabin décor of the dining hall, the photographs of dog sledge teams and wind-shredded Union Jacks, the dart board and pool table. The Americans are living in the present, if not the future, while Base R lives in the stalled Edwardian fable of Scott's Antarctic.

After his talk, Tom and I went for a walk around the point. At North Cove we passed patches of brown goo and blood, shockingly bright on the snow. The Weddell seals had recently given birth. I lingered. The chromatic deprivation of the Antarctic was such that any colour, even the brown of seal intestine muck, was salve for the famished eye.

"The trick of the Antarctic is knowing when to leave," Tom said. "You need to leave on a high." He would leave, flying out in the Dash, in a month's time. In a place where people voluntarily incarcerated themselves for two-and-a-half years, a month didn't seem very long.

"I know," I replied. "I hope I don't overstay my welcome."

"You'll see something very few people in the world will ever see, the onset of Antarctic winter." He did not sound encouraging, despite his words. Perhaps that was to be expected: the onset of winter in Antarctica was sobering, I was sure, a force of nature, like watching someone die.

I watched Tom pick his way amongst the stones. A new, pensive note had settled inside him. I didn't know him well enough to ask what was wrong. On the surface of it, the Antarctic encouraged sudden intimacies. In the Ellsworths I was alone (apart from Oddvar, Eric, and other boiler suits) in the world, more truly alone than I had ever been in my life, with Tom as my only companion. And yet something within me held back from asking personal questions.

Tom and I walked, observing the Weddell seals; as we suspected, many of the cows were calving. Their pups lolled by their mothers' sides, dewy lozenges of fur and fat.

Suddenly the sky darkened. Night was beginning to claim more of our day. Once a banished entity, it was returning with eerie speed. The sun lowered itself to a point just above the horizon. As the sun skirmished with the land its light was refracted into two sharp rays held apart from each other at a ninety degree angle, like the hands of a compass. One spear pointed northeast, the other southwest.

Tom and I climbed up to the monument to the dead at the Cross to find Gavin and Ben already there, staring into the sunset. We stood in silence, our bare hands shoved into our jackets.

It will never be any better. This thought came to me. What do you mean? I asked the voice that had generated it. I could feel a pressure coming toward me, distant but distinct, a barometric gloom.

Should memory be a shrine of moments? Certain moments in the Antarctic consecrated themselves in my memory. There they are, still, intact and recollectable, rotating like grounded icebergs.

This is one of them: Tom's pleasant, square face lit by the cold glow, Gavin's restless eyes scanning the sea ice. Alexander Island, Jenny Island, Ryder Bay, the icebergs still trapped in last year's sea ice in North Cove, Reptile Ridge, the string of mountains that stand sentinel, Valkyrie-like, behind Piñero Island. Behind it, Graham Land, the long spine of ice mountains tentacling north to Tierra del Fuego. The sun never sets, but neither does it appear. I will always struggle to describe what we see in that moment. I will write *bruises, lava, combustion.* Also *apricot, tangerine, carmine.*

If I close my eyes, now, I see us even now, four people at the end of the world bathed in the glow of perpetual sunset: Gavin, his explosive, eager intelligence; Ben, his shock of hair gelled by cold; Tom, the intuitive, dextrous pilot. Me, the writer/spy who will one day make too much of this, trying to capture the fugitive moment, which does not want to be held or known. The sun that never slept, and us, standing at the Cross, frozen in that moment of cold awe.

PART THREE
WINTERING

1. IF NOT, WINTER

rough ice
First-year ice subjected to fracturing and hummocking at the stage of young ice that has formed as a result of the freezing together of pancake ice or of fragments of fresh ridges.

FEBRUARY 16TH

In two days the air unit will begin to leave, flying the planes in pairs and in stages north to Canada. The Dash 7 will be the last to depart. It will go in early March, taking the essential summer personnel with it. The rest of us will be collected by the ship, the *Ernest Shackleton*, due at the end of March.

The end of March is six weeks away, but it is presented to us by Simon and Steward as if it is tomorrow. We have to start the end-of-summer clean-up, equipment has to be brought back from Base K and Ice Blue. Some of the winterers have already started packing. Their future is rushing toward them now, returning from a place where it has been held at bay. The Antarctic is one of the few places on earth where you can put your life into stall mode, in pursuit of science or money, or simply time out from the world. Until now, when our departure date is set, we haven't felt the press of the future, or even believed in its possibility. But like night it has returned. Now that it is back I feel the relentlessness of it. The future builds itself, moment after moment. It just keeps coming.

The summer camp atmosphere on base was replaced by an end-of-term feeling. The field parties had all come in and returned to the UK. That weekend Steward and the field assistants would fly in to close Base K and Ice Blue. Each Dash flight north to Stanley had a full cargo of passengers.

On one of those empty afternoons Gavin invited me to visit the marine labs. I had poked my head in before and knew what awaited: a supercooled vault filled with large, shallow tanks. In them lived the strange and fascinating species of the shelf and sea floor of the continent.

The lab was a new, purpose-built structure. It was built in some haste a few years before to replace the previous science lab, which had burnt down. It smelled like all the buildings on base — a dormitory/Portakabin odour I would recognize instantly for years to come, in university classrooms and temporary offices.

Gavin took me on a tour of the shallow aquaria. I peered into the jade waters. Among Gavin's subjects were giant sea sponges, some of them two meters tall; a man could crawl inside them. I looked at the giant starfish, giant sea spiders.

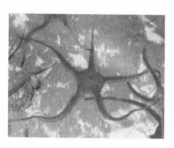

Gavin reached into one of the tanks and extracted a creature, a black disc the size of my hand. I took it. It was surprisingly heavy. I watched its tentacles wrap themselves around my wrist.

"It's a sea spider."

"Why do these creatures grow to such a size?"

"It's a paradox, isn't it? You'd think that in water this cold they'd barely survive, and instead we find these gigantisms. These waters are the most nutrient-rich on earth, despite the cold. There's abundant krill, and a lack of natural predators. Everything — sponges, sea squirts, starfish, spiders — flourishes here. People think that not much will be lost if these waters warm because they're barren of life. But things flourish in cold in a way they never do in heat."

Things flourish in cold in a way they never do in heat. Another sentence, casually uttered, that careened around in my head long after it was spoken.

"What are you studying here?"

In the split second before he answered, Gavin's posture changed. He stood straighter — he was at least six feet tall — and he seemed to expand: his lecturer posture, perhaps, or the version of Gavin who was often interviewed on television to "explain" global warming.

"The waters around the peninsula are warming faster than any other ocean on earth. Warming disrupts these creatures' nervous systems, and they start to die in large numbers. That in turn disrupts the food chain. All the creatures here are adapted to extreme cold. A single degree of warming and the krill could diminish, or even disappear. And then, well, it's game over for everything. Even the penguins."

The Antarctic is a tough and fragile environment at once, and these seemingly opposing qualities are inextricably linked. Once the creatures Gavin studied had adapted themselves to a relatively narrow range of normally killing temperatures, they made themselves vulnerable to even slight changes in their environment. Warming — normally a condition for abundance — is what will kill them.

I shivered involuntarily. I was bundled up in three fleeces, but it was not enough against the ambient cold of the aquaria.

"Let's get a cup of tea and have a seat in the office," Gavin offered. His office had a view of the runway. It was quiet, now that the Twin Otters spent most of their time in the hangar, undergoing maintenance for their long trip north. The sky was dull and overcast — a default Antarctic sky. Still, it had presence, a tangible dimension. There the sky was never a distant diorama, as in London. After my time in the polar regions, the milky tones of the temperate zone skies would always look fatally complacent.

Issues of *New Scientist* were scattered on the table in front of us. Their covers advertised articles about power blackouts, cyber-terrorism, aging populations, the psychology of suicide bombers, the dangers of materialism.

Gavin caught me glancing at them. "It doesn't affect us here, what's happening in the rest of the world: that's one reason why people enjoy coming here. It's a holiday from the real world. From threat. Ironic, don't you think, considering this place killed so many people who set foot on it, once."

Another paradox was also true, I responded — of being in a place whose destruction we were forced to consider daily, as well as our role in it, but where at the same time we were insulated from the negative capacities of civilization, from its rapaciousness, its trivia and bad faith.

"But the more I think about it, I see the Antarctic like a mirror, although angled permanently toward the future," I said. "In the rest of the world, we have to deal with certain threats. But here we're standing on a continent whose disintegration has the potential to threaten most of what we think of as civilization — coastal cities, infrastructure, agriculture. So actually we're at the epicentre of threat, although it's still in the future."

Gavin gave me a penetrating stare. He might have been deciding whether to tolerate this hazy panoramic vision, the writer's promiscuous horizon-scanning, unsubstantiated by facts or by computer models that calculate the future.

"This is the kind of place you can feel very abandoned when the planes leave," he said.

"Is that a warning?"

"Just a piece of friendly advice."

There was a strange note, empty, slightly clanging, to his voice. I tried to come to terms with his attitude: professorial, paternal almost, which grated and comforted at the same time. He knows more than you do, I reminded myself. Gavin dove under the ice for a living; he was ten years older and on *Newsnight* every other month, a world authority in his sphere. I was the ingénue: a necessary position for a writer, but almost never a comfortable one.

Normally I can't stand being the person who knows the least about a particular subject in the room, yet in coming to the Antarctic that was what I was, always. I was ill-informed in a place where scientific knowledge was the only authority.

"It will seem like a long time and also a short time," he said, referring to my remaining six weeks on base. Again I heard the note in his voice: a distant, wiry sound, like the call of an unfamiliar bird.

In the corridor between the laboratories, I bumped into Xavier. I was surprised to see him; I'd thought he would have disappeared north by now. Nearly all the scientists had left, ferried out on the plane to the Falklands. Base was small, but it could be surprisingly hard to keep track of where people were. Sometimes the departing scientists were so busy packing their instruments they didn't have a chance to say goodbye. The rest of us would run around base trying to find them to go skiing or watch a film only to find they left on the Dash three days before.

I hadn't realized the planes would leave so early. I had expected them to still be on base when the *Ernest Shackleton* came to collect us sometime in late March. But Tom told me that the flying conditions in the Antarctic became quickly unviable with the onset of winter.

"No, I'm staying until the last Dash out," Xavier said. "The prospect of a quiet place to write up the data is appealing. If I go back to Cambridge I'll be immediately distracted. I'm the line manager for fifteen people, et cetera."

"Do you like base life?"

"Base is a microcosm of British society."

"That's not quite an answer, Xavier," I hazarded.

"Well, I find it interesting. Here, you have something unusual —" He rubbed his fingers together, as if he were sifting sand. "Here you have the grain of experience. Things stick with you, like grit. Every conversation, every action. It's very . . . intense." His expression brightened. "It's easier, for me, not being fully English. I can be a watcher, here."

"Have you ever overwintered?"

"I did a couple of winters, a long time ago."

"What was it like?"

"It was —" He looked out the window of Lab 7; the day was overcast again. A steel current had entered the light in the last few days. "Long."

"I couldn't do it."

"Well hopefully you won't have to."

Winter is the real season: this was a mantra of the Antarctic. People meant the mystique, of course: the darkness, the starry isolation, the uncanny aurora borealis australis. The cold. On the other hand, something of a polar class system had developed over the years: those who had made it through the marathon of isolation were considered the elite of the Antarctic. There was a truth in this: such a voluntary stranding was tough, if often financially rewarding, thanks to the tax-free salary and wintering allowance they were paid. Winterers saw incredible natural phenomena, went on demanding winter camping trips; they had access to a knowledge that was off-limits to most of the human population.

As winter approached on Base R, I began to understand that summer had been a dream, a reprieve. Until so recently base had lived under the onslaught of a relentless sun. Now the sun's declination reapproached earth. On February 23 my shadow returned. I went for a walk around the point and found that I was being followed. I started and looked around. I had last seen this black simulacrum somewhere in the Falklands.

Later I passed Xavier's office and caught sight of him in front of his laptop, his greying hair, a white cable-knit jumper, surrounded by printouts of graphs. His computer screen was stuffed with coding.

I thought again how the glaciologists were a breed apart. The proficiency required with math and physics meant they were by necessity intelligent, but there was also an alpha quality about them — they were achievers, driven perhaps by the prescience bequeathed to their science by the climate-change agenda. They were also characters; they had passions outside of expedition kayaking and wildlife documentaries, precise and cultured obsessions you might sooner find among philosophers: tango, opera, jazz.

Xavier had given me one of his papers to read on the retreat of the Fennoscandian sheet. I understood about twenty-five percent of it, until I arrived at the math.

To look at an ice sheet is to regard something fatally static, or so you think. The ridges, palisades, and buttresses are charged with holding up the air. The icescape appears eternal. But it is in motion all the time. I learned about the violent pressures which develop in the base of the ice sheet, the heat brewed by the intersection of ice with rock, even while the surface appears implacable.

Xavier looked up from his screen. "Let's get a cup of bad coffee."

"No, I don't want to disturb," I said. "Besides I've had three cups already." Activities on base were winding down. No longer did we have evening lectures or group film screenings in the bar. Now all we did for entertainment was to sit in the dining room and talk, play cards, leaf through climbing magazines or *National Geographic*s we had looked at thirty times before.

We ended up going for a coffee anyway. We settled into an empty dining room, our coffees made pale by constipation-inducing Nido powered milk.

"So, have you found the inspiration you were looking for here?" Xavier asked. Everyone spoke like this now, as if our time in base was over, although I had five more weeks to run.

"I don't know. I'll only know when I start to write."

"Is writing always like that?"

"Always. Like groping in the dark."

He was silent for a while. "I'm not sure that writers — that creative people — realize that nature is purposeless. It is a pattern of unruly energy. You project the human onto it."

"I think writers are well aware of that, and that's why they write books. We have that in common with scientists. We study patterns."

"What patterns?"

"Aesthetics — style and so on — is actually about looking for the meaning that lies just beneath the surface of so-called reality. That surface is a decoy. To excavate it you need to use metaphor and stylistics."

"But you're not looking for information?"

"Yes," I admitted. "But in the abstract. What the writer is really trying to do is to give order to experience. To humanize life."

At this he scowled. "To humanize *life?*"

"Life is pretty inhuman, really. It's a disassociated state that may or may not support your existence."

"So what is the task for you here?"

"You can't write a book without characters. Literature isn't based on facts or setting but on society and on human conflict and choice. There has to be some moral dilemma, to make a story resonate," I said. "That's the difficulty — classic explorer narrative has one story only, the struggle against death. Which tends to drown out more delicate dilemmas. What you get in most Antarctic literature are monologues, soliloquies, barren stories of individuals' struggle to survive. We have a great literature of first-hand accounts, but otherwise fantasy, science fiction."

"But everything is science fiction. All this —" Xavier swept his arm in an expansive gesture, taking in the dining room, the flags, the heroic photos, the whiteboard where our fates and weather and transport arrangements were written "— is just *here*. It just is. The only truth is in the lived moment. There is no interpretation. There is only information."

"So for you things have a valueless reality, they just are?"

Xavier gave me a look very like the one Oddvar delivered to me in the melon hut at Ice Blue — a flinty, penetrating scepticism. "That is what I would like to believe. But yes, I think there must be something else, something beyond the facts. I've been unhappy, at times, as a scientist. There is a missing dimension to pure rationality."

He spoke in lithe cadences ruffled by his Indian accent that gave the golden ring of certainty to everything he said. Gavin also had this prophetic allure, but his authority was built on clarity and an obvious surface explosiveness. Xavier's intelligence was a rumbling one — remote, vaguely threatening, like a capped volcano.

"Before I met you I never realized glaciology was so overwhelmingly white," I said.

"Well it is. India has many glaciers and mountains, toward the north, so there's a fair few of us. But it's an eccentric interest, ice. At least in my country."

"What does your family make of you being a glaciologist?"

"In my family, you become a doctor, a solicitor, a politician. That's the limit of eccentricity allowed. But a glaciologist." He shook his head. "As long as I am successful and prosperous, they tolerate it. But I think they're resigned to losing me down a crevasse someday."

Xavier and I were the last to leave the dining hall, which emptied quickly after meals now. We were down to fifty-six people on base, near half the number present in those heady early days of January. There was an air of waiting: waiting to go, or waiting to stay.

I had a different office, as Lab 7 was required by the generator mechanic to talk to his family on Skype. My new office, Lab 5, was plastered with photographs of Wales, the Cairngorms, rugged British coastlines. I tried not to look at these photographs. They instilled in me a thin yearning, and behind this, a vague note of panic. I had so rarely been in a situation in recent years from which I could not escape. The simple fact of being confined to an isolated and far-flung place ought to have put in perspective the things I valued, ought to have provoked avowals about what I never would voluntarily eschew again. I began to write a list: *newspapers, espresso, broadband. Try to be serious*, I told myself.

I couldn't face head-on the possibility, which had only recently entered the picture, that our departure might not be so different from our by-the-skin-of-our-teeth arrival, that something unforeseen could enter the picture and delay, or even prevent, us from leaving the continent.

I looked out the window of my new office, toward the tiny cove favoured by elephant seals. As far as I could tell, elephant seals' day consisted of lolling, attempting to turn over and sometimes actually turning over, sighing, making half-hearted gestures at fighting (males), and emitting strange mewing sounds (females). I watched them until the sun slung low in the sky, then disappeared into the sea.

FEBRUARY 25TH

Tom and I go for a walk around the point. The day is cloudless and the sun so strong it stings our bare hands. All around us, ice is succumbing to its power, although Tom says this will be the last big melt of the season.

We sit down on the rocks. The bay glistens. Around us are sounds of fizzing and popping.

"When I first started flying up in the Arctic, twenty-five years ago now, I never thought I was looking at a place that would change. It just seemed so — monumental," Tom says. "Then hardly any of the science was about global warming. There was some resource conservation, some ecology. But mostly it was glaciology, traditional stuff. The idea was that the ice was going nowhere fast."

As he speaks, I feel such admiration for Tom — for what he does, his diction, his cast of mind. The admiration is fierce, a warm diffuse haze with a ribbon of chill threaded through it. The collision of hot and cold is not very different from desire. You could easily mistake one for the other.

"How long do you think you'll keep doing this?"

"I don't know. Sometimes I think it's my last season, that I can't take it anymore. Sometimes I think I could do it forever. Every year I wonder if I'm tempting fate by doing another season. That there's only so long you can go before something gives way."

"What do you mean, gives way?"

"Luck, I guess. Or judgment. Or both."

An explosion reverberates through the gelatin-still air. We watch as a mid-sized berg not more than fifteen metres away flips over in the water, snow and ice shearing from its edge. A blue-eyed shag that had been sitting on the berg takes flight and we track its arrow-like thrust through the sky, its long, sinuous neck straining, as if searching for something.

"I love to watch those shags fly," Tom said. "They're so committed. I try to fly like that, like an arrow thrust through the air."

Suddenly everything seemed to be happening in slow motion — the crack of the iceberg, the bird's flight. They stalled and hovered. Tom's words too. *I love to watch them fly. They're so committed.* My head buzzed as if I had just picked up a signal from one of the HF radio transmitters around the point.

I remember thinking, This is not an accident, or random, any of it: the blue throb of the sky, the exploding berg and, most of all, Tom's words. I have been delivered here, to this moment, for a reason. The connections between us, the iceberg, the mountain, the blue-eyed shag, appeared like a gossamer invisible ribbon, binding us, although loosely. But instead of looking at some mathematical matrix I was suddenly looking *into* things. Once inside, emanating from this vision was an awareness of the vastness of everything. Then suddenly I shot up into the highest layer of the atmosphere, and beyond, into space, from the starless Antarctic sky. Strange entities were there, huge and transparent. They knew what was happening, and what is supposed to happen, and what will happen. They knew everything.

This journey took a moment, too fast for my brain to record. I knew only that a high-voltage current of knowledge had surged through my mind. And then it was gone.

"I think I lost you there."

For a moment, I considered telling Tom about the brash ecstasy I had just passed through, or which had passed through me.

"I was just thinking," I said.

He nodded. He had also done a lot of thinking, I could tell.

I had never had a vision of the kind that had just seized me. I had always been unconvinced about God. It was too soon for me to share what could easily just have been a synaptic blip, a moment of reverie.

Although Tom might have understood. It was impossible to remain in the Antarctic for any time and not have to reset your interior horizon, that measure of our relationship to the earth and what might lay beyond it, not unlike the artificial horizon Tom and Lanier had taught me to read so attentively when flying.

The first blizzards of winter arrived. By late February my body expected more light, and instead it was getting less. We were losing half an hour of light a day. At this rate, I thought, it would be mostly dark by April.

My hair went limp, my nails stopped growing. Somewhere inside me, an alarm was building. I looked at the sun with a ripe, bodily fear. The calendar said it was spring and yet the sun was going away. I recognized an atavistic dread, one which said, The sun has burnt itself out, the harvest will fail, we will all die.

I toyed with the idea of making a joke out of all this, telling people in the bar about my weird fear, just because the days were shortening. But I noticed mine was not the only hunted face.

One afternoon I watched Xavier packing up his crates. Before I knew it, I was telling him about my experience around the point.

He stopped packing and sat down on one of his silver crates. "I think these things are not uncommon in the Antarctic — strange dreams, hallucinations, you name it," he said. "Here you're less protected — well, you're not protected at all — from whatever energies are coming down from the cosmos."

"You really believe that, as a scientist?"

"To an extent it *is* scientific: there's less ozone, the atmosphere is thinner, there's the tug downward of the earth's magnetic shield. I think we pick all of this up on a subconscious level."

The formality of his speech, the foreignness of his accent, was somehow more present. I had the sense he was measuring his remarks — not against me, but against something he himself had known, in his life.

"Or it was a genuine moment of enlightenment," he said.

"Have you ever had one?"

He was silent for a while before answering. The laboratory machines buzzed loudly in our ears. That was another effect of winter — the silence had become more silent.

"I never saw snow until I was fifteen and in England. I could have gone to the Himalayas, of course. I could have gone to the Hindu Kush. But I never did. I developed my fascination for snow, mountains, ice, in the north. I wasn't even particularly strong at chemistry or math. But I knew I had to master them."

Along the way, Xavier told me, he had been called into the Army, where he was an engineer. He had, all told, six university degrees, "although two of them are honorary." His father was a postcolonial administrator. His mother was a housewife with an unseemly interest in the occult. "I would come home and the living room would be full of shamans," he said.

"What did you think of your mother's interest in these things?"

"Oh, I couldn't bear it. It was all darkness and shadows. It makes you nervous, that realm, because you never know what's real. It's unhealthy as well as fraudulent. You go peeking behind the scenes of life and find there's some gigantic puppet master manipulating us all. You find there are dark energies at the core of our lives. And that bad things lurk in the future. How does that help you? My mother was anxious all the time, because she knew the future, or thought she did. My refuge was science, and rationality. I still believe the emotions are a dark force to be mastered by the intellect. I'm a throwback that way, I suppose."

We were silent for a moment. With Xavier I could speak my own language of ambivalence, of abstraction. For months I had been speaking another tongue, one built of facts and enthusiasm. I never fully mastered it.

"I see you've struck up a friendship with Tom," he said, after a while.

"He's a good talker. For a pilot."

"Great pilot too," Xavier said. "A real natural. It's necessary to have someone you can really talk to here. Otherwise it becomes unbearable."

FEBRUARY 28TH
If Not, Winter
No pain.

I want
I yearn and seek after
You burn me.

— "If Not, Winter," Sappho, translated by Anne Carson

Sappho's poems exist only in terse fragments. I think of the Swedish adventurer Andrée's ballooning expedition in 1897 to try to reach Canada from Svalbard, how they crashed on the pack ice after only two days. Their diaries, found in the 1930s when their camp on an outlying island of the archipelago was discovered, damaged and fragmented. Their last entries were a kind of polar Sapphic composition, with lines of ink erased mid-sentence by frozen ink or water. No one knows what killed Andrée and his two companions, but it seems they died within days of their final entries.

Scott's and Andrée's are the only two diaries I know of in polar literature which were left behind in the field of their endeavours and outlived their authors, but perhaps there are more. Diaries have an immediacy and poignancy, especially when you know they have not consented to expire but came to an abrupt end not of the author's choosing. The empty space after the last entries resonates into the future. Perhaps the future is just that: more and more white space.

As the reality of winter made itself felt, we were given new instructions by Simon, the base commander. We were not to walk around the point without telling him first. The sea ice was hardening at the edges of the shore, and with fresh snowfalls it could become difficult to see where the land stopped and the sea began. We risked falling through.

Safety was taken very seriously on base, and indeed in my whole time with BAS I felt safer than I ever have in my life. On the ship we had regular safety drills and major incident exercises. On base the fire alarm shrieked weekly and we would have to file outside to the veranda. We had endless runway and aircraft safety drills (which boiled down to: don't walk into the propeller); then there was the very thorough mountaineering training we were put through by BAS as soon as we arrived on base. In all these rehearsals it was drilled into us how in the Antarctic tragedies had more than one cause; it was almost always a series of mishaps, or misjudgments, which sealed people's fate.

How to think about safety in a rescueless place? In cities at least, we live in a regulated, regimented habitus, micromanaging our environments, swaddled in health and safety legislation. This gives rise to a dangerous delusion that we are in control of our lives. By nature I veer between extreme caution and recklessness, without any middle ground in between. The Antarctic suited me on that score.

It was also a place of manic jests, of innocent risks. Like the day, in the name of fun as well as safety protocol, we immersed ourselves in zero-degree Antarctic water to test the immersion suits (they worked fine). Or the day we spent on Léonie Island having a friendly conversation with several large male fur seals (which could deliver a mean bite and which we were under instructions not to approach) after they came to sit right beside us and all but drank from our cups of tea. Never mind what we did on field trips in the planes, flying so low sometimes over the snowfield for a good view we risked snagging the landing gear on sastrugi.

I had become less, not more, concerned for my personal safety. After all, our progress was being followed by an organization that would make a strenuous effort to save us, should anything happen. But beyond that I can't really explain the raw serenity I felt.

Tom felt the same. We talked about this on one of our trips, in the cockpit, as we were approaching base again, flying over Alexander Island from the south. "You know, I never think of crashing," he said. "I mean, we could go down, sure. But I always think I could get the plane on the deck. We wouldn't disintegrate."

"I know," I said, "I feel it too." That somehow we were protected.

That evening I went to ask permission from Simon to go for a walk around the point on my own. I found him in his office in front of his computer. Beside him a paperback edition of *The Count of Monte Cristo* lay face down. Many people on base set themselves reading-marathon challenges, tackling the stout nineteenth-century classics: *War and Peace*, *A Tale of Two Cities*.

"Just remember what I said and stay away from the edges," he said. "If you fall in walk back to base as quickly as you can. Here, take this." He handed me a radio for good measure.

These walks were perplexing. I always started feeling a relief to be away from base, which meant away from other people. But forty minutes later I would round the head of the point glad to see its olive-coloured buildings again.

I stopped midway, trying to spot where the old path had been. It was now covered with snowfall. I ought to know the way well enough by now. I looked out over the bay to the east. The silence washed through me. In January, I found it soothing. Now there was something ominous about it. Something was nudging me. It was strangely external, not an intuition or instinct. I couldn't hear what it was saying: *Go home. Go back to the world*, maybe. I had one last chance to leave with the last of the summer staff, Xavier among them. I knew there was an empty seat on the last Dash out. Otherwise there was no way out until the ship in early April.

I looked out into the waters of Ryder Bay to a by-now-familiar scene: rose-gold clouds, a horizon studded with mountains, icebergs the size of office buildings.

What would I be doing, though, at home in London, 14,500 kilometres away as the crow flies? I tried to picture myself going to a film at the South Bank, shopping at H&M, cycling to my Clerkenwell office. I felt it like a blow, the absence of my friends. For months I had been too busy to miss them. In the meantime, they had gone on with their lives.

My breaths refused to make puffs in the air — there was not enough moisture for it to properly condense. I stood looking out to the barren flanks of Piñero Island. To the north the sky had turned graphite. A storm was on its way.

I wait in La Vie en Rose to meet him again. It is early July and thirty-five degrees. I feel like ripping the soupy heat aside as you would a curtain, to get beyond it, to a realm where thoughts have clarity and purpose.

I have written my exams, I have graduated. In a delayed and useless emotional response, I am frightened. If I have not done well on my exams, I will never escape this place. Not only my future, but my very life, depends on escape. Somehow I have known this all along, but did I work hard enough? Have I engineered a way out of this trap?

Six years in this town have taken their toll, have transferred something of its world view. Would it be so bad to be driving a Chevrolet Chevette through its snow-lined streets to be a teller at the Bank of Montreal, or to teach in the very high school I have just exited, or to be a secretary in the English department in one of the two universities which repose on the flanks of a sloping hill?

There are thousands of towns like this all over North America, and probably other countries too: curiously humourless towns with their one bad-boy nightclub, the unofficial stripper joint on the edge of town, the strip malls — Sobeys, Canadian Tire, Atlantic Superstore. The convenience stores and opticians and the weepy clapboard mansions.

All through high school the message has been clear: don't stand out too much, don't be too smart or too different. Don't be a freak. The characters the town holds worthy of its gaze are lawyers or hockey players for men, while for women all that is required is that you are beautiful. Certainly there are characters too: local poets and cosmopolitan CBC radio hosts and eccentric philosophy professors and hippie entrepreneurs like Elin, drawn to the raw wilderness that forms a cordon around the place, to the autumn apple-picking festivals and the spring horse shows and sleigh rides through shadowed forests in winter.

But underneath the tightly patrolled skin of the town, a rebellion slithers. Trailer park violence in the nearby Army base accommodation, Masonic gatherings above the furrier's shop — the same place our society of unofficial bohemians holds weekly screenings of avant-garde French films — and of course the attacks on women.

I want to get out of this place untainted, but by what? Not only that delusion it brews, that here you have reached the apogee of life. *Toronto and Montreal are stressful. Have you seen the traffic? In other countries people kill each other, just like that!* No, a larger threat is its glutinous energy that seems intent on sucking the life force out of you, with its gleaming cars that purr down avenues in slow motion. I feel I am moving through gelatin; the air itself has given up and elected entropy.

"Would you like another coffee while you wait?" The waitress has a note of sympathy in her eye. She thinks I am being stood up. I decline. I am still in the grip of the dream I had the night before, about the river, the man.

In the dream there was a man who was my father, or stepfather — the dream-man was neither the man I wait for now nor Mark nor my grandfather, but an unknown man with coal eyes and dark hair, thin and vital, in his mid-forties perhaps.

In the dream he was my father but there was something awry about the relationship; it was not a normal father-daughter scenario. He was very angry with me, he felt I'd betrayed him. Something had gone terribly wrong between us. But he did love me. I could feel it. In the dream I killed myself because of it, or he let me drown in the river. We were in the water in any case. It looked like a giant river, too big for this country, even — the Amazon, perhaps. He held me in his arms.

Passion. I could feel it in his arms, his being, like a force field, a fanatical energy. I have never encountered or inspired such an emotion in my waking life. But I recognized it in the dream as electric, a current to live your life by. It would be a life lived like an elongated dream. An obsession.

He is now very late. An alarm begins to tug at the edges of my skin. A kind of premonition.

I feel certain the dream I had the night before was important. The dream has left me with a certainty I can't articulate to anyone. It is too spooky.

I feel I will meet my father again in the future. By then I will be someone else, and he possibly will be too. I will be older than him, maybe. But we will know each other still and our hearts will thud against our chests painfully, with a spasm of recognition. We will confuse what we feel with fear and we will want to hurt each other. We will not understand where this instinct comes from, or what we once were to each other. I have read about reincarnation. I feel the power, the possibility, of it now in my life. How each time we come back into this realm our previous memories have been erased, and we start somewhere new, with the innocent faith required to face life, with all its hazard and pain. We start again, as someone else.

"You never call me by my name," he says. He has arrived, at last. His mouth has that disappointed curve I will see again, many years later, on the mouth of the man I will call Loki.

"I don't like names," I say.

We are eating ice cream. He has ordered caramel crunch. After a couple of mouthfuls, I shove the bowl away.

"What's the matter with your ice cream?"

"It's too sweet. You never call me by my name, either," I say.

"Name taboos," he says. "In some cultures you can never address your sister-in-law by name, for example."

"Why not?"

"Because . . . it's usually about . . ." he stalls.

He manages an effortless change in topic. He jumps from one moment to another without hesitation, like crossing a river on a shattered causeway of stones.

Years later I will study anthropology and realize what kind of sandbar he had nearly stranded himself on. Name taboos exist between in-laws, often. You cannot use your sister-in-law's name, because by naming her you might want her, you will give voice to our instinct to desire the forbidden. They are about controlling desire; often they are a specific bulwark against incest.

After our meetings in those long blue evenings, I take overly circuitous routes home. I find myself walking alone for blocks and blocks, my internal rear-view mirror invisible but positioned.

Girls pass me, in twos and threes, clumping together for safety. How can he resist these teenaged girls, the killer, with their stringy legs, their neon green halter tops? Their perms and crimped hair in imitation of Pat Benatar, Paula Abdul. They've been watching MuchMusic from distant Toronto, trying to figure out what is in style out there. The purple nail polish like individual molten bruises on their toes. The scent of young flesh.

I am new enough in the world that I still struggle to separate the things in progress versus things which are definitely over. High school is over. My life in this town is nearly over, although I must wait two more months to be fully released. The summer days swirl into each other. I can only tell they are passing from the cobalt shadows they throw on the future, how, like the evenings, they lengthen, before diminishing.

2. NIGHT FLIGHT

hummock

A hillocky conglomeration of broken ice formed by pressure at the place of contact of the angle of one ice floe with another ice floe.

MARCH 5TH

Minus ten today. A different cold hovers on the fringes of the air now. Sharp, like needles, but unlike the cold at Ice Blue, a dampness curls around it, making it sink into our bones. I feel something inside me — some buried capacity — respond to this new calibre of cold. An old stiffening.

On the runway now the cold slices through my three fleeces. I feel a slight pain in my lungs. New oblique winds take swipes at me as I run back and forth from the top of the airstrip to Jenny Island. I find the hollow grandeur of the island oppressive, now. I don't know when this change took place, or what it means.

A snowstorm blanketed base. After so many years away from it I had become unused to cold. There is a vigour in cold, of course; it awakens, with its ethereal, negative energy. It feels easier to think in cold. Blizzards, on the other hand, force you inside yourself.

I watched as Christmas stencils formed on the windows of my office. I paced back and forth, bouncing between the Brecon Beacons (*Scenic Wales* calendar: May) and the flat, hot sands of the Gower (August).

Deep winter of the kind I experienced each year of my life in Canada always put me into a cryogenic slumber, emotionally, a state from which it felt impossible to emerge. Hunger, taste, desire, all sunk within themselves, cowed by the hostility of the outside world. This was one reason I haven't missed truly cold winters all these years in England: while I am alive I want to *feel* alive.

The day before, Tom and Lanier had flown up to Stanley to collect provisions for the coming month until the ship arrived. They had to return the same day because the weather was closing in.

"Is that common, to fly up and back in one day?" I asked Tom before they set off. I knew this meant ten hours' flying. We barely had ten hours' daylight on base now.

"It's knackering," Tom agreed. "But we've got to do it." They had a twenty-four-hour window to fly the last crew of non-essential personnel north, pick up supplies for winter, and return. The day after that they would leave for good, taking Steward the FOM and the remaining two air mechs with them.

I was still in bed when I heard the plane take off early in the morning. That afternoon, around five, the PNR siren sounded — the last Point of No Return of the season.

PNR was one of the Antarctic tropes which no one explained; like the bing-bong, or internal PA system on base, everyone expected it was self-evident. The Point of No Return is the place at which the plane no longer has enough fuel to return to the point of origin — say, the Falklands. It is committed to land no matter what the weather does. So if the weather worsened and the runway experienced whiteout conditions, or the cloud level suddenly rose above 12,000 feet, where ice crystals begin to form, the pilots would have to keep on coming to land at base, even if there was a danger the plane could ice up. Base had to be ready should an emergency landing be needed, so the PNR siren rang out over the base and the runway with its sharp, plaintive bleat.

A couple of years before, Tom had told me, the weather had "crapped out" — pilot-speak for zero visibility — and he and Lanier had to land the Dash with no visibility and no visual references from the landscape, relying on instruments but without the aircraft guidance systems used by airports in the rest of the world.

Everyone on base was mustered to put on a boiler suit and handed expired flares. They lined up at the end of the runway and, when the Dash was expected, lit the flares over their heads, in a human runway light beacon.

In the dining room we all listened to the siren. "That's it, folks," Steward announced. "The last PNR of the season." Now so much of life on base was about the rituals of leaving and ending, looming abstinences: *the last banana, the last cucumber.* Even for those of us who were not staying for the winter, it felt as if we were about to walk the plank.

About half an hour before the Dash was due, I went with Darren, one of the St. Helenan cleaners, to watch for the plane's arrival on the veranda of the accommodation block.

Darren was one of a triumvirate of young men from the island who worked each year as domestics on base. The Saints, as they were called, were related. They were gentle-humoured; I never saw them look anything other than content. They had beatific conferences by the juice machine, conversing in a rollicking island patois. They did not join in with the fancy dress parties and quiz nights, but we could hear them talking and playing cards in their pit rooms every evening.

Darren stubbed out a cigarette. "Sorry," he said. "I'm so homesick now I need to smoke." He was not looking forward to the trip home on the *Shack*, as the *Ernest Shackleton* is called, which had a reputation for unpleasantness at sea. "I'm the kind of guy who gets seasick in the bathtub."

We scoured the sky for the four over-wing lights to appear somewhere over Jenny Island. We had never seen the Dash land at dusk before.

"Where are they?"

The runway lights were lit for the last time in the season; in two days' time they would be covered with old oil drums. Drifts would accumulate around them, creating obelisks, wind-scoops, cornices.

"There she is."

We saw four sparks of light under a black sky. Behind was snow-capped Jenny Island. The Dash emerged from the sky, flying slowly over the ice-choked bay. We watched it drop lower against a sunset of chilled silver. I felt a shiver go up my spine. Even though I had witnessed the pilots' skill many times by now, I still couldn't believe what they did. The Antarctic gave their exploits the tang of heroism; in this case flying up and back to the edge of civilization in a day, returning in near-dark to our slumbering colony.

That night Tom came to find me in my office. He sat down wearily, perching on the edge of the chair. Even flying the Dash for the short stints when I had been given the controls had exhausted me, and he and Lanier had just flown ten hours.

Everything had gone according to plan, he said. "Just a routine marathon."

The colour had drained from Tom's face. Soon he would be gone. He was perhaps my only friend left on base. His imminent departure had become a doppelgänger — something that looked exactly like Tom, but wasn't.

He excused himself and rose. "Well," he said. "Goodnight."

MARCH 8TH

A blizzard hits. The Twin Otters are twitching to get away but they can't. "Typical," Tom says.

The day is a blur of snow. We are not allowed outside, other than to go to our pit rooms or the laboratory. These we can find in the whiteout, out of habit. Otherwise we risk wandering off in the blizzard and becoming disoriented.

The news comes at sit rep. A weather window will open tomorrow. Tom and Lanier will fly the Dash north, taking the met man, Xavier, Steward, the air mechs, and ten others with them.

I couldn't sleep last night. All night I lay in my bunk bed, staring at the pink and grey swirl of the mattress above me. Base is so different. The shags, storm petrels, and elephant seals have moved on. The atmosphere between those of us who are left is thickening. There is an astringent current within it, like milk threaded by lemon juice. It might sour.

The day and night before the Dash left I struggled with my instinct to be on the plane. A voice I may not have heard before but which lived inside me began to whisper, then, more urgently, to speak aloud until it was almost a shout: *Go*.

I had asked to stay until the last ship out because I wanted to witness the finality of the closing down of summer. I felt sure that this scenario would play a part in the novel I would write based on this experience. But was there another reason — was this an obscure game I was playing with myself, to test my resolve, or even to test fate? Did I want to put myself in a rescueless place to see how I would respond?

People had been left to overwinter unexpectedly in the Antarctic before, although not recently. If something went wrong with the ship, I didn't know what would happen to the twenty-one of us who were hoping to resume our lives in the English spring. The planes certainly wouldn't come to get us. They would be in Canada by then, undergoing summer maintenance. I already knew that Base R was scheduled for a late uplift, and that for various reasons no other Antarctic organization was extracting its personnel from so far down in the continent so late in the season, when the sea ice would be thickening. There might not be any other ships in the area to extract us.

On Tom's last day, I showed him my two most recent books. He had been asking to see them.

He flipped through them, pages and pages of accumulated words. "I don't understand how you can sit down and write all this," he said, finally. "Make it all up. I'm sorry, but I just don't get it."

"I don't either," I admitted.

"What drives you to do it?"

"Because I can, I suppose. What drives you to fly?"

"Because I love it. It's the power it gives me. I don't think I could be an ordinary person."

"People who don't know how to fly are ordinary?"

"No, not exactly. I meant I didn't want an ordinary life. I wanted to see things few other people have seen before. I wanted to live on my wits and my skill."

"It's the same for me."

But as I said these words, I wondered. To be a writer in the Antarctic was an even more ridiculous occupation than usual. There, everyone could do something of immediate use — navigate a RIB in the ice-choked waters of the bay, plot a course on a nautical chart, dive beneath the ice and identify twenty species of sea cucumber, fix a diesel generator or manage air traffic control or fix a Twin Otter or study UVB-resistant Antarctic moss.

Base society was suffused with a particular elixir — so many people there emitted a force field of personal power, constructed, as in Tom's case, from years of technique, practice and experience. They had been military logistics officers, they had been pilots in the RAF, they had sledged across the ice field for days, they had spent months in cold ice domes coaxing ice tubes from the continent in order to know the planet's future. Base required you to trade on your personal expertise or, if that was not in the equation, on the power of your persona. Writers, on the other hand, are usually introverts; their personas are projected onto the page, in part I think to absolve themselves from the necessity of projecting a version of the self solely aimed at lodging itself in others' minds. You might only know who they are by their books. Otherwise they move through the world like ghosts.

"I don't know how you do what you do, either," I admitted. "When I think of what could go wrong —"

"I never think about it." Tom's voice was uncharacteristically stern. "Remember that. Never think about what could go wrong."

I wished I had been less conflicted in my life. I have been driven to be a writer in order to expiate these conflicts, which multiplied effortlessly.

"Sometimes I wish I didn't have to be what I am." As soon as I said this, I heard its adolescent ring. I knew he wouldn't tolerate it.

Sure enough, a flash of impatience moved across his face. I had always known Tom had the capacity to be dismissive, to have little truck with ambivalence. But it had never been directed at me before. "Well, then don't," he said.

He was impatient to leave base, I saw. His mind was elsewhere — on the blue miles of sky he had yet to traverse.

Along with Simon the base commander and Melissa the doctor, I was given the honour of being in the official sending-off party. We crossed the runway with the departing summer personnel. Everyone, including Xavier, had changed in the last twenty-four hours. They had an air of detachment, ready for their imminent re-entry into the real world.

Tom and Lanier went ahead some hours before to prepare the

Dash. They'd received the latest wind charts off the internet and their faces were creased with concentration. Conditions were okay, Tom had told me, although they could have been better. "Our only divert is Punta," he said. "Marsh is shut. Something military. The Chileans aren't telling us."

We waited on the apron as the passengers boarded. I could see Tom and Lanier in silhouette in the cockpit. They were moving around, taking checklists out, flipping dials. Eventually Tom reappeared wearing his distinctive peach-coloured pilot's Ventile jacket, that genius invention from the Second World War.

"Well, take care of yourself. Make the most of it and don't worry about anything. If it all gets too much, I'll come back to get you."

We laughed. We both knew this was impossible.

"I won't say goodbye," he said. "Pilot's superstition, I guess."

"Like the not looking back," I said.

"That's right. 'Don't look back, don't say goodbye.'" He looked down at his feet, kicked them together — for warmth, I suppose. "Well. I'll write from the road."

Tom went to join Lanier in the cockpit. The passengers were seated. Steward appeared in the doorway and hauled the stairs in, giving us a final crisp wave.

The runway was covered with a dusting of snow. Tom and Lanier turned over the plane's engines. One by one the propellers began to whip the air, creating eddies of gravel and flurries. From my time in the jumpseat I knew that they had the white laminated sheet of the departure checklist balanced on the controls and were working through it methodically: *spoilers retracted, avionics on, elevator trim set for takeoff.*

The plane trembled with the combined thrust of the four engines. It turned with that demure grace it had for such a large machine, and taxied to the end of the runway. The tremble became a shudder. The takeoff run was sudden and absolute — depending on its load and the wind the Dash could be airborne after six hundred metres of the runway's nine hundred metre length. When it passed us standing by the hangar, it was already flying.

Tom and Lanier banked the aircraft around the peninsula. They would return and fly over base for the traditional end-of-season flyby. For a while the aircraft disappeared from view. When it returned it was flying so close to the ground, gravel on the runway scattered.

As the plane headed for us, the drone of the engines was deafening. The pilots climbed and the fuselage passed above our heads. We could see its pale underbelly. Then Tom and Lanier climbed further to clear the icebergs at the end of the runway, and soon the plane was only a dot in a silver mesh of cloud.

Everyone drifted away, back to the warmth of the dining room, walking in slow clumps across a runway where suddenly there were no aircraft, and we would no longer have to obey the siren warning us not to cross. A sharp wind had brewed from the south, but I stayed behind even as it started to snow, staring at a place in the sky where a plane used to be.

MARCH 12TH

My thoughts are smudged suddenly. I can't write more than a few lines at a time. If I pick up one of the books I have brought with me for research — Robert Macfarlane's *Mountains of the Mind*, for instance — I can only read a paragraph.

I can just about manage to read email. Tom writes to me from the "ferry flight" as the pilots call the long journey to Canada. They must take a route that allows the Otters to refuel every eight hours. The Dash can fly longer; all the planes have extended range thanks to full auxiliary fuel tanks.

Before Tom left, I photocopied atlas pages and drew a thin red line between his stops: Stanley, Montevideo, Florianópolis, Rio de Janeiro, Recife. Then a long hop to Caracas, then Curaçao, the British Virgin Islands, then Texas and Montana. The final leg will take them to Calgary where the planes will be maintained over the northern hemisphere summer.

Tom had seen so much of the world from his cockpit capsule, and not only the ice world. I envied the marvels he was witnessing and which he described briefly in his emails — the giant thunderstorms of Rio Grande do Sul, a lavish sunset over the Amazon. "I'm just an observer," he'd told me on base. "The scientists, they're the ones who are doing something about it. I just watch it happen."

There always has to be a watcher. I suspected he was made for this, and that it was part of the reason why he had become a pilot. "They're loners, these pilots," Gavin had said, and it was true. Tom was self-sufficient. He only really needed himself and his plane, to fly through the air, to feel that power in his hands. Everything else was supplementary. I had the sense women were like this for him, too, the equivalent of his new flat, new motorcycle, new car — each one catching his interest for as long as it remained unsullied by fault, shimmering with transitory beauty, not so different from the land passing by underneath him: seas, glaciers, mountains, a wide, lazy arc of rivers.

For a couple of days after the planes left, I felt calm despite the eraser-like quality of my thoughts that deleted themselves as soon as I perceived them. I tried to get on with my work, making notes, consulting the library.

It first happened at lunch in the dining room. I was eating tinned asparagus soup. The anxiety did not build slowly. Like the Dash on its takeoff run, it went from nought to sixty. Knives erupted from everywhere in my body: my lungs, my mind.

Some creature, valiant but black-hearted, had leapt on me from behind. Spooked by the ferocity of the strange attack, I retreated to Lab 5 and to the photographs of green, beaches, sunsets. Loneliness washed over me. Without Tom's or Xavier's presence, the insulating safety of their comradeship, I was exposed.

You've made a mistake, the voice inside me said. You should have been on that plane.

Who are you? I asked it.

I decided I should be very busy. First I went to the sledge store and learned how to wax skis. In the mechanics' workshop, I held a greasy skidoo chain while one of the wintering mechanics replaced its drum. In the alcohol store, we stacked and re-stacked boxes; the facilities manager showed me how to use the Miracle Span recycling machine. On Fridays there was "scrub out," when we cleaned in a messianic frenzy, collecting fluff-less wine corks and dust-less debris from corners.

Winter's peculiar marriage of industry and contemplation overtook us. Winter is a fallow season. If we were to lose winter to a warming world we would forfeit this time of dormancy, reflection, soul-searching. We would lose the starved grandeur of this season of ice and snow, as well as the tenacity it confers upon those who last its course.

The strongest characters are forged in times of little hope. Without winter, will we become flaccid, sybaritic; will we become perpetual heliotropes, those flowers that turn their faces to follow the sun?

MARCH 14TH

Sunday is the longest day on base. Now we are forty-two people; twenty-one winterers and the rest of us who are leaving on the *Shack*. Now people come to breakfast late or not at all. Or they come in their pyjamas, as if we are children at a sleepover. They seek to re-create a Sunday "back home" and sit with the two-colour printouts of the online version of the *Observer* we must all share, swapping sections when we are done.

Later, there will be a Sunday lunch. *What are you doing today?* we ask each other. We can ski, or go back to bed to watch videos, lying underneath the duvet. Later in the bar there will be games of backgammon, pool. Then a film night, pints of Guinness.

The day is dark; our first taste of real winter. The windsock says thirty-five knots, maybe gusting to forty, from the northeast — an unusual wind direction. Later I will go for a walk, but only if the wind subsides. A strong wind can augment the wind chill to a point where your face will freeze, even covered in a balaclava.

I work with the nightshade pulled down against the blizzard outside, and concentrate on the cheering miniature universe I've tried to create around my desk: photographs taken with the underwater camera of the slope over by Lille Island. Starfish, sea squirts, sea sponges — flares of colour in the deep.

That afternoon I did a head count of those of us left on base. Now we came in twos, like Noah's Ark: two boatmen, two carpenters/builders, two chefs, two vehicle mechanics, two base commanders, two meteorologists, two engineers (mainly meteorological, one outgoing, one incoming), two plumbers, two doctors (one outgoing, one incoming), two terrestrial biologists, two marine assistants, two marine biologists, two electricians, two guys whose jobs are unknown to me. And then the unsettling odd numbers, the singles and the threes: one facilities manager, one engineer, one station support manager, one operations assistant, five GAs, one base GA (the "PermaFID" who can't keep away from the Antarctic), three Saints, one writer.

Now that we were a small group of people, the distrust and suspicion of the writer ripened. I had more cover when base was full of us exotics: journalists, government ministers, foreign office people, Royal Navy trainees. To counter my growing estrangement

I became conspicuously jolly, chatting in the dining room for hours to people I'd never spoken to when there had been a hundred people on base. It didn't work. I knew most people thought me a spy of sorts. But they needn't have worried: I wouldn't write about them personally, in a way they could be recognized. To write about the essential emotional truths I witnessed and experienced, I would need to create entirely different people. The only entities who required no such protection were the Antarctic, and me.

MARCH 15TH

The skies are our television. The light is slanting and elusive; it changes from second to second. We see whipped meringue clouds, as if some giant had raked a five-pronged fork through them, separating them into five perfectly rutted architectural seams. The clouds mushroom until they hover overhead, consuming the sky. Then there are the lenticular clouds — high velocity mountain clouds. These are frequently taken for UFOs. They do a good impression, these slim saucers that peak at the top. We begin to see nacreous clouds too: they waver like thin sheets of mother of pearl. The colour of green melons, the subdued silver of the moon.

My anxiety sharpened. I was no longer treating it like a transitory symptom; now I was frightened. That day I went into the rarely used bathroom by the doctors' surgery to throw up.

Opposite the bathroom, old topographical maps of the peninsula were taped to the wall. In close scale they showed our immediate surroundings in Graham Land: the Oscar II, Foyn, Loubet, Bowman, and Fallières coasts, the Eternity Range, the Trinity Peninsula. The dead men/sci-fi names chilled me.

The taste of bile and spit stayed in my mouth. It was hitting me forcefully that there really was no way out of here until the ship came, the knowledge a greasy shear in my stomach, my mouth, my mind. On the tagging board on the way back to my office I looked at Tom's plastic nametag. It hung in the *Flying Off Base* section, waiting for him to pick it up at the beginning of the following summer, seven months in the future.

Over the next few days, the anxiety and insomnia worsened until I could only sleep a couple of hours a night. It felt as if I had imbibed a truckload of coffee. Every muscle clenched. The panic was prefabricated; it had been manufactured somewhere else in my life and shipped to me intact and posthaste. I was aware that it was an abstract panic, of finding myself outside time and abandoned, but I couldn't reason it away.

I would manage to get to sleep around two a.m., but would consistently wake soon after to the hiss of snow against the window. We were losing nearly an hour of daylight each day. I would look at my watch every evening as I sat in Lab 5, trying to write. Now dark at seven thirty p.m., now at seven ten; the next day, six forty p.m.

I began to take sleeping pills but would waken two or three times, woozy. When I closed my eyes I saw shimmering white figures against the screen of my eyelids: white wolves, white faces with elongated faces and slit eyes — Steven Spielberg–film aliens — and figures I could only describe as Vikings, helmeted, with braided hair in pigtails. All of these creatures would look at me disapprovingly, as if I was not up to a mystery task they had set me. They could have been right. I couldn't eat, I couldn't sleep, I couldn't wait for time to pass and time, in retaliation, slowed down and stopped, then, with an almost audible grind of gears, went into reverse.

I told myself that it was the spooky speed at which the light was going that made me see these things. I remembered Xavier saying that strange visions were common in the Antarctic: ultrarational physicists had them, JCB drivers had them. Why shouldn't I?

Even the atmosphere felt different now that I was thinking of such things, as if all the energies of the planet had sunk and were congealing in a cold layer of thought. Scientifically there is some evidence for this: the earth's magnetic field is drawn downwards by the weight of the ice continent and distorted, so that magnetically the planet looks less like a sphere than a pear.

New threats blossomed in my mind; as soon as one was banished another took its place. I imagined war, epidemics, the collapse of the global economy, tsunamis. Other threats were more proximate. There was only one ship to take us out, and the Drake Passage at that time of year was notoriously rough. I realized belatedly that by staying I had put myself in a situation where there were no guarantees.

One morning in sit rep, Simon informed us that all power would be shut off at eight p.m. All the power had to be turned off to simulate a failure of the main generator over winter, and the backup generator must be tested in its ability to handle the basic workings of the base.

I sat in my office awaiting the darkness, drinking the last of the bottle of Sangre de Toro Tom had given me before he left.

The generator powered down. Suddenly there was no light. We were not allowed to light candles on base — the wood and air were so dry we could easily spark a fire. I put on my parka and went outside. It might be my only chance to see base in total darkness.

The sky was stark with stars. Their light was cold, more platinum than silver. Silver-cobalt clouds filled the sky.

Others joined me. We looked into a little-known quadrant of space. The sky was stitched together by constellations I was beginning to recognize; waypoints like the Southern Cross tilted on the horizon, soon to disappear. The night sky pulsed and flared. The stars had more intensity, also volatility. They blinked like frozen nebulae, a dying pulsar. I felt vertigo knowing we were looking down into space. Antarctic skies seen by very few people, only by telescopes trained to find the echoes of the Big Bang, by satellites and space stations.

Around us the world was hardening. In the bay it started as an impasse; days of ice flowers, tiny pellucid formations, accumulated in the shapes of crystals. As the carpet of flowers knit together in the cold nights they were soaked by seawater and transformed into grey gruel; overnight, as the temperature plummeted, the gruel became porridge. Within days ivory pancake ice formed. Ice welding itself together produced a metallic sound, like steel grinding.

Simon stood beside me. We watched the sea ice in the gathering darkness.

"This is it," he said, staring into the darkness. "The sea ice has begun to form. This is the beginning of winter."

Donna's house is beside the cement factory her father owns. He built the house himself. It is cream-coloured, with pillars at the entrance. All the floors have thick carpet, apart from the kitchen. On the kitchen wall is a clock in the shape of a butterfly. The hands are converging on midnight.

It's hot — over twenty degrees still. Tomorrow will be a real scorcher.

Donna and I are sitting at the kitchen table. I am sleeping over that night.

"I can't imagine what it would be like to meet my father now," she says.

There are things I can't imagine either. I have never imagined myself as a lover, although I know of other girls my age who are definitely some boy's, some man's, lover. The word itself threatens to disintegrate every time I try to hold it in my head or it mutates into its close cousins. Loser, loner, lover — only one consonant separates them.

My friends seem to think that being someone's lover is only a matter of time, a choice we or someone else will make, but I'm not so sure. I have always felt more mineral than human. I think of the coal in the Carboniferous forests, edges pressed and pressed by centuries of decay until I am as smooth as the soapstone seals sold in Shades of Light. Would anyone want to enact such a fleshy, trembling manoeuvre with me?

Michael comes in. The screen door creaks behind him. His eyes are red around the rims. As he makes himself a cup of instant coffee, I see his hand shaking very slightly.

"What have you been up to?" Donna asks. "Tonight wasn't your patrol night."

He sits down beside us. "What have you two been doing?"

We glance at each other. Not that we have been doing anything particularly illicit. We've spent the evening grooming Donna's quarter horse, trying on new shades of eyeshadow, watching television.

"Talking," I say.

"About what?"

Donna sighs. "School, guys, killers on the loose. What else is there to talk about?"

He sits back in his chair, takes his baseball cap off. His hair is plastered to his forehead. He runs his hand through it.

"You've been driving around looking for him, haven't you?" Donna says.

Michael looks off into the distance. " I just thought I'd let him know that we're watching."

"Where did you go?"

"Along the river. The woodlot roads. The usual places."

"Did you see anything?" I say.

"A couple of winos sitting on the park bench, the two that are always there."

He opens his mouth, closes it. "I talked to my lieutenant today. He said we should set a trap. Give him the situation he likes best, and wait to catch him. A woman, a girl, alone. Maybe her car broke down and she has to walk to the payphone. Maybe — I don't know. Some situation where she's out there alone, and then we wait for him."

"You'd get a policewoman to do that," Donna says. "Go undercover."

"There isn't anyone on the force young enough. All the victims have been in their teens or early twenties. The youngest we've got is a thirty-year-old. And she's fat."

I find I have said it before I have thought it. "I'll do it."

"No you won't."

"Why not? Besides," I say, "if something happens, you'll save me. Right?"

He looks at me again. He has heard the note of challenge in my voice. There is a glint in his eye this time, sharp.

In that town darkness does not distribute itself evenly, but gathers in clumps like bunches of blackberries. The woods are glutted with blackberries and blueberries, thickets of mauve corpuscles — I have learned the word from my science teacher; this is what lungs are made of. He showed us a slide on the overhead projector. They looked like the dusky purple grapes shipped from Ontario or Florida.

The road is a woodlot road, just up from the old bridge with green arches that groan in the night. Pulp trucks rattle across it, their loads unstable, like the one that killed the girl at my school.

I slap my thigh and bring my palm away bearing blood. The mosquitoes and the blackflies will eat me alive, before the killer ever gets to me.

Don't tell anyone about this. Michael's words ring in my ears.

Michael and his partner, Dave, are out there, somewhere, in the patrol car.

The walkie-talkie is flat and cold against my skin. I wear it concealed between the small of my back and bottom, inside my jeans. My outfit is ridiculous: jeans, pumps, a tank top, turquoise — the kind of clothes I would never wear. I have make-up on. My hair is crimped — Donna did it.

I walk along the gravel shoulder of the highway. Cars pass, beep their horns. The sound continues long after the car has gone by, like a ragged ribbon flapping behind. Two boys, their faces yellow streaks, yell out the window, "*Yee haw!*"

I watch the cars, waiting for the red taillights to flash on. The vehicle slowing down, waiting for me to catch up with them. Or maybe even backing up, pulling up alongside. *What's the matter, darlin'? Need a ride?*

I walk in the cooling empty air, the crunch of gravel beneath my feet. Every so often a lumber truck thunders by, and I am nearly knocked off my feet from the air it displaces. Its balsam smell trails long behind it. Little pieces of bark fall off and get caught in my hair.

After an hour of walking, no one has tried to kill me. Finally I hear a car pull up, a spew of gravel as it eases onto the hard shoulder. I turn around.

"Hey," Michael says. "Time to call it a night. Get in."

We repeat this scene, four, five times on different roads next to the river. It is an experiment that Michael and his partner have hidden from their superiors. He is hungry for promotion, I suppose, hungry to be the hero of the town. As for me, I am used to walking on the sides of roads. I have spent several evenings the past winter walking thirteen kilometres home from the stables in the dark.

There are no more attacks that month. It is as if the killer knows I am a decoy and has decided to take temporary retirement while this piece of backwoods theatre is played out for his benefit. I feel certain there is a relationship of some kind between our efforts to ensnare him and his reluctance. An understanding, a truce.

My mother comes into my room the next morning. I am worried she will see the unfamiliar clothes spread on the floor. But she doesn't seem to notice them.

"How is it going?" she asks.

"What do you mean?"

"With — him."

Him — she refers to him as I do, evading the name and role neither of us can bring ourselves to say.

"Why do you want to know?"

"Because." An unusual expression alights and at once flees my mother's face.

"What's the matter?" I press.

"You can't always trust that what he tells you is true."

"What do you mean?"

"He tells lies," she snaps.

I laugh.

"What's so funny?"

"I thought you were going to say something much more serious."

"You don't think lying is serious?"

"Then why did you encourage me to meet him now?"

"To get him off my back. And because I thought he could have changed. Maybe he's had counselling, or whatever."

Doubt sprouts in my mind. "What kind of lies?"

"Anything," my mother shakes her head. "Everything. I didn't realize it myself, until it was too late."

Michael and I sit in the police station, in his office. We are alone — all his colleagues are out on patrol. We have come in from another failed decoy session. I sit there, ludicrous in my tight-fitting jeans and neon green halter top.

"You're different."

I narrow my eyes. "What do you mean, different?"

"You're hard to get to."

"Hard to get *at*, you mean."

"To reach." Michael draws back. His expression says, Are you happy now? Just as with my mother, the main currency between us is frustration. "It's as if you've decided everyone is out to get you."

"Everyone is out to get me." As evidence I gesture to my outfit, our surroundings. "Why would you want to reach me?"

"It's natural for us to want to connect, don't you think?"

"Why? Because I'm your sister's friend?"

"What do you want?" he asks.

"I just want more control."

"Over what?"

"Over what happens." I add, for emphasis, "to me. I just want a say here. A say in my life. That's all I want."

He seems to consider this for a while, his expression at once bemused and sombre, as if it is a fabulously exotic request. Then he plays his trump card. "I saw you at the Dairy Queen. With a man. Older."

I don't want to tell him but I will need it, as a defence. "He's my father."

Michael raises an eyebrow, triggering a bunching of skin on his forehead. He is only twenty-three. Those lines hadn't been there before. His job, I suppose.

"We're getting to know each other. He's from out west."

"Well that can only be a good thing." He sounds older then, knowing.

I study his face — a face I have known since junior high, or maybe even earlier. I can't remember when I first met Donna and started to visit their house, packing away my *oh my Gods*, and *for God's sakes* on their doorstep in deference to their strict Pentecostalism, donning my friend-who-goes-to-church persona like a smock. Michael was always there. Always seven years older, always watching me, even if he seemed to be ignoring me — playing basketball, driving off in his first car, blowing a kiss to his sister. It didn't surprise me at all when he joined the police force. He had good peripheral vision.

"We're going to try again, Thursday night."

"I'll be there," I say.

Those dusk evenings of summer when I walk down woodlot roads on my own, waiting for the feel of hands around my neck, nails piercing skin, a blow to the head — the murderer has used all these tactics — I recite the names of the trees of the province to keep me company. For my university entrance exam I wrote an essay on them, quoting their genus and species in Linnaean classification, their distinguishing characteristics, their susceptibilities to the cyclical plagues of insects that beset the Maritime seaboard.

They rotate in my mind, an internal compass in verbal form, like the psalms my mother intones during Mass: cedar, pine, balsam fir, balsam poplar, oak, maple, walnut, Siberian and European elm, larch, hawthorn, laurel, alder and pussy willow, sumac, sycamore, white spruce, black spruce, balsam fir, jack pine, white birch, trembling aspen, tamarack, beech, maple, black walnut, hickory, oak.

My mother is folding laundry, her face softening at the small white jumpsuit her youngest baby has just outgrown.

"I think he honestly doesn't know the difference," she says. "He thinks he creates his own reality, so they're not lies. He thinks if he says he's a pilot, then he's a pilot."

"Are you sure he's lying about that?"

My mother purses her lips. "Who knows. He might actually be a pilot." She laughs again, that tinkling laugh, like so many stones dropped on a marble table. I will hear this tone of voice many times in my life, perhaps I will even voice it myself — that of a woman burned by her own susceptibility.

Yes, she had avoided that particular fate, of being with a man who is perhaps a fantasist, but perhaps not. The stories he'd told me were detailed — the make of plane and model, its capability, wingspan, engine thrust, the length of the runways, fuel capacity, weather. The exact sort of detail Tom would later recount to me in the Antarctic.

3. CLIMBING

cirque

A large circular or nearly circular steep-walled recess or hollow in the side of a mountain or a hill, generally ascribed to glacial erosion.

MARCH 18TH

It's been snowing for days. The sat dome is coated with, as the comms manager says, "the wrong kind of snow." It is slushy and cancels our comms. We are deployed to brush it off, standing on ladders in a bitter gale. Now, stepping outside is like entering a blasted open-air cathedral, the organ wind grinding some nameless requiem. Simon the BC tries to jolly us along: "Big deal," he says, "it's only minus ten. Real winter temperatures are minus thirty."

George the radio operator and I decide to go climbing. It might cure us of our cabin fever. We are both twitchy. For several days we wait for a weather window. Finally it comes — the forecast goes up on the whiteboard: Sunday will be "bonza!"

The day was blue and clear. Our destination, Stork Peak, stood on the other side of the glacier, just out of sight of base. One of the highest points on Adelaide Island, Stork Peak looks out onto the Wormald Ice Piedmont toward Laubeuf Fjord to the east; to the west is open sea all the way to Australia.

We put on our cross-country skis and skied up the ramp to Vals. From up above base it was another hour's skiing across a flat ice field. We were weighed down by ice stakes, an ice axe, climbing ropes, and jingly janglies.

The snow squeaked; the only other sounds were our breathing, wind, the scour of snow as it was harried by the wind. It sounded like distant snakes hissing.

When Stork was in sight, George and I took off our skis and roped ourselves together to cross the crevasse field that lay at the bottom of the bergschrund. It is better to cross crevasses on skis, as they distribute weight evenly and make it less likely you will fall through the bridge of the crevasse, which might be as shallow as only a foot or two of snow.

The sky was rigid with sun. We needed the strong contrast to pick out the crevasses, which can be near invisible; when you can see them they appear as fractures drawn in a very slightly lighter white than the surrounding glacier.

George went first, prodding in front with a stick, the climbing rope taut between us. After a few steps the stick failed to meet resistance and passed clean through. He looked over his shoulder and pointed downward.

When I reached the place where his ski pole went through, my instinct was to tiptoe. During the time it took me to pass over the crevasse, I could not stop myself from imagining the ground giving way beneath me, the glass-hard edges of the slot, then the fall into the crystal chamber.

The deepest and longest crevasses on earth are found in the Antarctic. They tend to occur in patterns: intersecting, chevron fields of zig-zagging lines, or long fractures parallel to each other. Sometimes they are visible from the surface but very often they aren't — you only know they are there when you find yourself falling. In Antarctica crossing a crevasse is the equivalent of crossing the road in London. You do it often, whether you are aware of it or not. Mostly the mouths of crevasses are gummed together with snow, and this snow bridge may hold up to forty-five to fifty-five kilograms. But you never know.

The crevasse field behind us, we switched our ropes and re-knotted them for climbing. We took our crampons, ice axes, and snow stakes from our backpacks and these, along with the weight of the rope around my neck and the karabiners chinking on my waist, meant that I was hauling half my weight again up the mountain. I called to George, who was climbing dextrously above me. We took a break, although I could tell he was eager to crack on. It took us an hour to reach the summit.

There I struggled to stand upright. After an hour of climbing, my thighs trembled.

We stared in silence at the view: mountain, ice field, mountain, ice field. Wave after wave of blue and phosphorescent white ploughing into the horizon. There was Orca — so named because it looks like a killer whale's black dorsal fin, the sister peak to Stork — and Trident. Beyond these, we could see the edge of Adelaide Island and the ocean.

We breathed in, our breaths condensing on our eyebrows and eyelashes, frosting them. George had been there before, but I'd never stood on the summit of any peak before, let alone one in the Antarctic. I understood the appeal of mountaineering at last: we hadn't walked to a destination, but into a dimension.

In *The Spiritual History of Ice*, Eric Wilson quotes Wordsworth's assertion that we need not literally climb mountains to undergo what he referred to as cosmological experiences, but that we "can attain such heights *psychically*." In book six of his *Prelude*, published in 1850, even as he fails to summit an Alpine peak, Wordsworth's pilgrim "apprehends the transcendent powers of his imagination," Wilson writes. "He realizes an unseen, infinite power, a universal mind animating and organizing" the entire world — the fearsome peaks as well as the placid valleys. He realizes that home lies not in sensual attachment to matter but "with infinitude."

As Wilson argues, quoting Wordsworth, Wordsworth's hero becomes "something ever more about to be": "a mountain theologian, a *limen* between sky and earth, boundless reaches and rugged forms." *Limen* means threshold, as in the limit of a

physical or psychological response, or the place between space and form, spirit and matter — the pattern of invisible immensities. Wordsworth's poetry suggests that we are in a circle "whose centre is everywhere, and its circumference nowhere," to quote Shelley. Mountain climbing is a metaphor for climbing toward God, toward the infinite. Here, in the thin air of fear and endeavour, we are closer to the divine.

The ice fields of the Wormald Piedmont stretched into the distance, until they ran into the broken wall of the Shambles glacier, its snow-waves arrested at the moment of cresting, like the frozen tsunami Max told me about on the ship.

The air temperature freeze-dried my sweat instantly, but the sun kept us warm. We sat in the snow to eat a bar of ten-year-old Bournville chocolate — "quite new, for the Antarctic," as George said.

As we ate, George and I talked of Good Ice Years and Bad Ice Years. A Good Ice Year is when you can travel long distances over the ice. Through his binoculars we could see far beyond the bay, out to Piñero and Pourquoi Pas islands. There, refuge huts had been set up thirty years ago, in case sledging parties had been caught on the sea ice. "They're full of thirty-year-old manfood." George said. "That makes ten-year-old chocolate an appealing prospect."

Between us and the huts were fifty kilometres of crevasses, capricious sea ice, fast-running channels. If we had to reach those huts on sea ice from where we stood, we would face ice cliffs and crevasses galore. We were on the eastern flank of Adelaide Island, on the Wright Peninsula, only a speck of the Antarctic compared to what Scott, Shackleton, and Amundsen had traversed, and even such a relatively short journey seemed an impossibility.

George told then that two men from Base R had been killed in 1981 on the Shambles Glacier, not far from where we stood and only a few hours' ski from base.

"What happened?"

"They went on a jolly, climbing like we are now," he said. "On the way back they fell into a crevasse."

I thought of the story Elliott told me at Lockroy, of the French yachtsman who went for a short walk and died. Or the Twin Otter operated by a Canadian charter company that crashed into an iceberg after takeoff from Base R ten years before. Apparently a passenger survived the initial impact, but he had a broken leg, and the base rescue team could not safely traverse the ice in the bay to reach the iceberg in time, and the survivor died of exposure.

Around us fine crystals of snow were scurried by the wind. This faint rustling was the only sound. But then I detected a vague buzzing, like the distant sound of an aircraft. I had heard this once before in the Ellsworths, on our thrilling mission in early January. It was the sound of me, a humid hum of blood pumping, the electrical field of a body, normally drowned out by the sounds of everyday life.

The sun was overhead but soon it would dip. It was time to descend. George and I picked up our packs. I took one last look at the scene.

Even though I had been in the Antarctic for months I was still shocked by the landscape, its sudden violence — black nunataks, the omnipresent crevasses — blended with satisfyingly iterative patterns of stern vertiginous mountains rippling into the horizon, thick icing cakes of unstable ice, ripe for avalanche, and the pale blue seams of crevasses and bergschrunds that ribbon across them, like stripes in white candy. It was so impervious to our gaze and our touch that its rebuff felt personal.

This is another moment I tried to seal against time: George and me sitting in the snow on the top of Stork, looking out into the frozen world of early winter Adelaide Island, eating a bar of time-warp bitter chocolate that looks no different from the day it was manufactured. Everything shimmers. Far from implacable and flat, the ice field seems to be vibrating. Or maybe it is just the beating of our hearts.

The days darkened. Our skies became lambent and unbelievable. New icebergs arrived in the bay, envoys of winter. Simon told me they were remnants of the Pine Island glacier. An enormous girded berg had drifted in on the gyre from the Bellingshausen Sea and sat in the mouth of the bay. On the water grease ice formed, its surface a dull grey matte that neither absorbed nor reflected the little light that penetrated our dingy skies.

The runway was frozen. Running was exhausting, as I had to break the hard outer layer of snow, then try to extract my feet from its soft underbelly. The wind chill brought the temperature down to minus thirty-five. I didn't need a thermometer. I knew this threshold from experience, the stinging-tightening feel that meant my skin was freezing, so I rubbed my face as I ran and kept my fingers moving constantly. There was not much I could do about my toes — running in mukluks would be even more exhausting. I didn't sweat; my body wanted to, but any moisture was instantly freeze-dried, even beneath three layers of fleece. Icicles built up on my eyebrows until I was running with a miniature ice-ledge on my face.

Fur seals had colonized one end of the runway. There they lolled and barked and mated. The males were territorial. A male and his girlfriends wallowed in a moat of snow beside the airstrip. I would try to step gingerly around the harem, running in a wide arc into the middle of the strip, but the males growled and gave chase anyway, slumping along behind me, barking, like some computer-animated hybrid dog-seal. Meanwhile the north end of the runway was claimed by outraged mother skuas who dive-bombed me. I ran with the shadows of these miniature pterodactyls swirling around my head, flailing my arms in a polar version of *The Birds*.

People watching in the cafeteria got a panoramic view of a snow-covered, panting Frosty the Snowman figure, bundled in fleece and a couple of hats, chased by the palace guard seal-dogs, then set upon by the pterodactyls. The field assistants told me my daily attempts at keeping fit were the only real amusement left.

What they didn't know was that these daily routines of exhaustion were the only way I could keep myself steady. More than the anti-anxiety pills I had started to take, even, the runway was my saviour. There, shuttling between the fur seals and the pterodactyls, I tried to dissipate my anxiety and strip myself of the cloying, perverse cloak of Antarctic Base Time.

The week before I had finally taken the decision to do something. The anxiety attacks, if that is what they were, had spooked me. I was now mixing my anti-anxiety pills with antidepressants — quite a feat for a pharmaphobe who usually hesitates to take an Aspirin. The antidepressants made me feel as if my edges were blurred. The physiological effects were obvious — strange chills rippling through my cerebellum, the instant insomnia — but the emotional ones were harder to pinpoint. Antidepressants often seem to make people feel contained, as if their emotions have been herded into a holding pattern. The pills mute the feeling, fraught with dread, that you are going to lose composure. But in turn they dull the lineaments of the self, leaving an unfamiliar, if calm, husk.

I hadn't had an anxiety attack in twenty years. These episodes were different from the ones I remembered having when I first moved to Britain. In the Antarctic, the fear seemed to be generated by my body rather than my psyche. I might have been suffering from claustrophobia and feelings of abandonment, but surely these were not strong enough to produce vomiting and insomnia, the sprinting accelerations of my heartbeat, the blaring, red-eyed fear. I asked myself — or my body — What's wrong? The answer came back. I don't know. Just get me out of here.

The telephone rooms were barren of all decoration, apart from an ice axe affixed to the wall; Very convenient, I couldn't help thinking, for when the Base R resident gets the Dear John phone call from his girlfriend, or learns that his father has died suddenly and she will not be going to the funeral. I kept a leery eye on the ice axe as I telephoned friends, in Rio de Janeiro, in Sussex, in London, in Toronto. Within a few minutes of talking to anyone I was in tears. I couldn't understand how this was happening. I called them only to have a conversation, but the sound of their voices, so far away, in another world, turned my panic up several notches.

As winter settled, the landscape turned dark and blank. I no longer woke up expecting or hoping to see sun or blue sky. The continent was closing in on itself. Each winter the Antarctic doubles its mass in sea ice, effectively locking itself away from the world for seven months as the planet tilts away from the sun.

My belief that something would go very wrong intensified. I was convinced the ship — which had not yet left the Falklands — would sink and I would be left in the Antarctic for a winter, or forever, among people I didn't understand and who didn't like me. I spotted a small item in the world news section of our weekly printout of the *Observer*: "Buenos Aires engages in sabre-rattling over islands." Argentina had suddenly slapped regulations on shipping, requiring any ship flying the Falklands flag — as the *JCR* does — to remain outside of Argentine coastal waters or face seizure.

Other fears were new to me. I didn't know where they came from. I was assailed by sudden if vague memories of exile, of being separated from my family, in a time before even letters. There was no communication and I died without seeing them again. But when? Whose memories were these, since none of this had happened to me? These fears didn't belong to my life, so what was their providence, some other life?

The paranoia might have been generated by the chemical tension of the pills I was taking for the anxiety. It felt as if all my brain cells were shredding themselves. The pain was on a neuronal level. I clenched my teeth, had heart palpitations, bright lights flashed across my mind — garish fairground colours, purples and oranges.

I had read about Arctic hysteria — *perleroneq* in Greenlandic — which comes on the back of the sudden autumn darkness. In the Arctic it affects not only people, but dogs; this reassured me, that not only the weak and fragile human psyche succumbed. The afflicted — people and dogs — often erupt into violence, or fits. Sometimes they try to kill people.

The despair was spiritual, mental, and emotional. Something was so wrong with me, it was like a disease. It was both part of me — generated by me — but external.

I had come up against an adversary I had never met before, and it lived inside me. It seemed I was experiencing a direct confrontation with those darkling energies that course just under the skin of our existences, so hidden and elusive they might be like shy animals in the night: once you go looking for them, turning your flashlight on those hidden dark corners, they take fright and run away.

MARCH 23RD

We fill our days with routine: breakfast, smoko, the Team Timed Crossword, lunch, smoko, dinner, then blockbuster films in the bar. There is an air of exhaustion; at night in the bar, few people talk with animation. Most stare silently into their cans of Guinness. We are almost out of beer and ready to start on the cider stocks — a sobering prospect.

I try to read. But I find I can't read anything about the Antarctic. But all my books are about the Antarctic. I try to write. This is worse. I break out into a cold sweat.

I instruct myself to sit down. I deal myself a few basic facts: You are here to write a novel. You will never be here again. You need to live this life before you can write about it. But you also need to get going. Novels do not write themselves.

Class dismissed, I pace and try to free myself from this nameless vice grip by thinking. An idea begins to cohere. I start on the book I have been sent here to write.

To write fiction requires a subjunctive cast of mind — what if something *were* to be true, then how *would* things be? It is the *what if* approach. For example, what if, in an alternate Britain in the late 1980s, cloning science advanced sufficiently to have a race of clones living alongside "real" human beings, so that their organs could be mined to keep their legitimate brethren alive? That is the scenario that informs Kazuo Ishiguro's great novel *Never Let Me Go*.

The *what if* that presented itself to me was less sinister, but it had a similar dystopian tinge. What if, a few years in the future, a pandemic destabilized the world, and in the chaos that ensued a researcher, a writer much like me, were left to overwinter? What if a young woman had died on base a few years before, in mysterious circumstances that looked like suicide?

Before I went to Antarctica I thought, Whatever happens in the novel I write out of this experience, no one is going to die. I'd had enough of the equation between the Antarctic and oblivion. But now, having lived on the continent for long enough, I knew how available death was, here, and how to die in the Antarctic would be an unordinary death. It would by definition be spiritual; in its white negation of human life, you would not so much die but dissolve into a historyless dimension, into particles of light.

As I sketched out this fictional scenario in my mind, I looked out of the window onto the runway. Now that the mountains surrounding base were covered with snow, if, for some emergency the pilots did miraculously return, they wouldn't be able to use their exposed flanks as markers. Everything was too white. When visibility crashed, you could easily find yourself in what the pilots called the "Ping-Pong ball": white sky, white land, and enveloping mist. "Only your instruments tell you which way is up and which is down," Tom once said.

But caught in the Ping-Pong ball, many pilots, even very experienced ones, cease to believe what their instruments tell them and listen to their inner ear, or allow their eyes to manufacture ghost horizons. The plane begins to tip in their hands, imperceptibly at first, but then at a greater and greater angle, until they are locked in a tight downward curve, the dreaded "dead-man's spiral."

I wondered if this was what I was in, emotionally. Now that the planes were definitively gone, I realized that our isolation was not only procedural, because there were no planes to airlift us out, but also symbolic. Flying is power, and knowledge. When we can no longer fly over the Antarctic the landscape assumes its textbook description: a threat, an ice jail, another planet.

That night a red-gold moon appeared in our sky — an autumn harvest moon. I stood under it in my shirtsleeves. Mark, a field assistant from Northern Island, appeared and tapped me on the shoulder.

I'd spoken to Mark several times. He had a thin face and watery green eyes. He was friendly, older than his cohorts, a little older than me, perhaps, and easy in his own skin.

"Aren't you a bit chilly?"

"I don't know."

"Come on, I'll show you some pictures of my trip to Berkner in January. I heard you wanted to get out to Berkner with Tom."

"Is that who you flew with?"

"Yes, he's great, Tom. He likes to show you things. It's almost as if the Antarctic is his home, and he's proud to show you around."

Mark's photographs appeared on the screen of his laptop. We saw the Ronne Ice Shelf depot, called Bluefields, the miraging sea smoke that rims it, produced by the open ocean coming smack up against the white cliff of the ice shelf. Mark told me about his three drop-in visits to Base Z, once to have a shower and do his washing. These visits were so swift he didn't even meet anyone on the platform base.

Base Z perches on the edge of a giant ice shelf 1,600 kilometres to the east of Base R. While it can accommodate up to seventy people in the summer, it has a core overwintering staff of only sixteen. Base Z is much more inaccessible than Base R. A ship, usually the *Shackleton*, does relief in late December, when the changeover of personnel takes place. Other than flying visits from Twin Otters, that is the sum of Base Z's interaction with the outside world. For 105 days at the height of the Antarctic winter, the sun does not rise above the horizon. And while on sunny summer days temperatures can rise to a balmy zero degrees Celsius, in the winter temperatures normally stay below minus twenty with extreme lows of around minus fifty-five.

Mark would be leaving with us on the ship.

I asked him, "If you had to stay another winter, how would you feel?"

"It wouldn't be the end of the world. There's enough food on base to last two years, probably more."

"Is there anything you miss about wintering?"

"The skies," he answered without hesitation. "The aurorae. You see skies down here you can see nowhere else on the planet, skies no other human being has seen, possibly, ever." He landed on the word, *ever*, with biting conviction.

"Also, I like that I haven't locked anything with a key or paid for anything in a year and a half. I wonder if I'll be one of those winterers who gets to the Falklands, goes into the supermarket, and walks off with a chocolate bar without paying for it."

Mark leaned toward me over the table and looked around, as if to make sure we were not being overheard. "It's so great to forget that we live in this exhausting world where *everything is for sale.*" He drew back. "There's one thing I'm looking forward to, though — seeing strangers. People I don't know. Faces. People I'll never talk to, even. I never knew that was so important."

"But the people who come down here in the summer, they're strangers — you don't know them for a while."

"But you do get to know them," he said. "When I first came here, I thought the limitation of the Antarctic was variety — there was only a set kind of food or set possibilities within a single day. But the limitation is actually possibility."

"The possibility of what?"

"Anything. You might meet a complete stranger, or do something unexpected. Whether it happens or not is unimportant. But the possibility needs to exist."

Mark had just handed me the key to my growing internal emergency. It wasn't about only our physical confinement, or the fact that there was now only one way out, but that every day was mapped before it began. There were a limited number of people, some of whom I had never spoken to at all, and would not, due to enmities and cliques and entropy. The possibilities of each day were constrained; I would go for a run outside or on the treadmill or for a walk around the point; I would read the books lined up on my desk and write up my handwritten notes and work on my book.

But, I asked myself, in my usual life in north London, would it be any different? Each day would similarly have a limited range of possibilities. I would go to work at my job or write at my desk, go to the supermarket, I would talk to my friends on the phone — it wasn't as if I would be delivered a life-changing chance encounter simply by occupying that position on the planet.

But in London there would be the potential, however remote, for something to happen, a chance meeting, an invitation, an event brewed by moving one's body through a space replete with buildings, streets, people, parks, galleries, restaurants, buses, and trains. I might see a person I would never see again, whom I would know nothing about, but it would be unanticipated. It would be a surprise.

Here the same forty-two people would eat breakfast in silence, heads in cereal bowls while leafing desultorily through a *National Geographic* we had looked at many times before. Then we would eat lunch together, then dinner.

Looking at Mark's photos lifted me out of my strange terror, if temporarily. I went back to my office. The window was plastered with snow stencils again. In front of it were my books, lined up in alphabetical order — in an effort to quell panic I had turned into an uncharacteristically ordered person, organizing my books, sorting and re-sorting notes and photographs. My computer files now had titles like *Magic, The Cosmology of Ice, Crystallography, Albedo, Climbing, Capitalism, Coleridge, Scrying, Catastrophism, Faust, Thomas Mann, Fridtjof Nansen, Edgar Allan Poe, The Little Ice Age (Europe, in the Middle Ages), The Last Glacial Maximum, Aliens.*

I left the building. Base was flooded with silvery blue moonlight, like foil. The ring of mountains around base reflected it, and the full moon on the snow was so bright it drowned out the starlight. I had wandered onto a stage set for a chill Wagnerian opera, the moonlit mountains a theatre backdrop for giants. They might have revealed themselves at any moment, clad in taffeta, magnanimous, beckoning us to a dimension of ease, of golden summers and bucolic trees.

Chills rippled through my body — from the antidepressants, likely, more so than the drama of the view. My brain felt tight, as if it had been placed in a vice.

I stood there, surveying the grandeur of the mountains, its crisp silence, until I felt the cold soak into my bones.

MARCH 27TH

Eight days to go before the ship arrives. It was due to arrive tomorrow but has been delayed in the Falklands with engine problems.

Time presses itself into my eyes. I struggle to open them in the morning, I struggle to keep them open. New visions have replaced the wolves. Flowers, black and canopy-like — miniature umbrellas — bloom in fields fertilized by flesh. Fungi flowers feeding on the rancid cadaver of hope. I close and open my eyes, willing the flowers away.

We have three hours of lilac light in the afternoon. Venus and Mars shine on the edge of the western horizon. The snow on the runway is too deep now; I've had to transfer my allegiance to the treadmill in the food store. I run for an hour, staring at boxes of tinned mushrooms and cauliflower, a vista so dull I miss the fur seals.

I go back to the office and stare at the runway, the line of black and red bamboo flags that lead up from what used to be the crossing point, curving up the traverse; the oil drums that cover the runway lights, a scoop of wind-drifted snow curling against them.

I am enveloped by a brass ringing sound, the sound of no sound at all. I know from my very basic physics that negatives form until they occupy just as much space as a positive force, if not more. That these forces can expand infinitely until they explode, creating a dark void of antimatter. This was how our universe was born — in this negative radiance of eternal darkness.

It feels like a group of cheetahs are consuming me. They gnaw and gnaw, they are growing faster, more lithe; crucially, they are luckier than me. They'll leave me for dead, then move on. Time has that smell, something amphibian and engrossing, bent on plucking you from the earth. It is much, much bigger than me and I am afraid.

The last time I ran on the runway, I found the North Cove end littered with the corpses of skua chicks. I remembered Xavier telling me about the bird's habitual fratricide. While two chicks are born, one kills the other, just as shark embryos devour the other in the womb.

"Filial love is a fiction," he said. "Think of how many children try to harm their younger sibling just after birth. They know very well what threat it presents to their survival, the arrival of a brother or sister. Think of *King Lear*."

"And is that the case in your family? Have you wanted to kill each other?"

"No." I could see he was shocked by my question. "My brothers and I, we love each other." He fixed me with that look again, poised between suspicion and interest. "So your book, the one you will write: will it be about survival?"

"I don't think so."

"All Antarctic stories are survival stories."

"Well maybe it's time that changed."

In my office I was surrounded by these books, survival stories all: *The Heart of the Antarctic*, *South*, *The Worst Journey in the World*, Scott's journals. They were appealing in their simplicity, I was coming to think. There was a threat, but it was obvious and external: the cold, starvation, exhaustion, the crevasse, the whiteout. There was the journey and the arrival, which was fleeting, anticlimactic, incomplete. When you look at that photograph of Scott's party at the pole, all facing the camera, swaddled in their inadequate clothing, their faces black with cold and exhaustion, you see not one note of barren triumph. Their faces are soot-streaked, frozen: awful masks of dread and failure. They look like they are thinking, *Let's just get this over with.*

I wondered if I really was meant to come here. Denise's strange vision of me standing in the snow dressed like an "Eskimo" was likely just a psychic stab in the dark, one that I took far too seriously.

But the idea that there is a crystal river of fate, of circumstance and consequence, is far from a crackpot idea believed in only by telephone psychics. Shelley, along with many Romantic thinkers and writers, was convinced that the crystal could throw light onto the future. In the Victorian era, scrying was the esoteric craze du jour.

In the crystal river, time flows forward but in it shapes congeal; if you learn to read these shapes you can see the general direction of the future. In the case of sentient beings such as humans, they "are determined to be moderately undetermined," writes Eric Wilson in *The Spiritual History of Ice*. "Some are able to gain greater distances than others from the flows of fate. These freer beings are thus able to entertain wider arrays of possibilities for turning these flows." And so, he says, man is capable of concocting "complex plans of navigation."

This seems to suggest that the direction of our lives might be dependent upon our psychic energy — meaning the flexibility of the psyche rather than anything purely esoteric — and not destiny. Fate itself might be determined, but if our directional energy is come-what-may, it might be possible for outcomes to remain undetermined.

Wilson writes,

> *The human mind yearning for holistic powers beyond the divisions of time and space perhaps recollects an origin in an eternal consciousness . . . if the divisions of space and time are*

illusory, based on arbitrary mental habits, it could follow that a mental principle beyond space and time generates life and being . . . if "to be is to be perceived," then an intuition of a universal mind is just as valid as an empirical perception of mechanism.

This may be the crux of the relationship between cosmic power and poetry, specifically of the Romantics so associated with snow and ice: Coleridge, Shelley, Byron. Wilson suggests that the poet's mind could be a glacier, creating and destroying as it flows with the universe of things in the "eternal rhythm of terror and tranquility." Could ice be a bridge between mind and matter? Ice — no longer an agent of death, the manifestation of a coming frozen apocalypse, but rather of scintillating revelations.

We are hopeful creatures. I had believed that everything would be all right. I believed in myself, in my future, in my luck. But in the Antarctic, for the first time in my life, I began to think such a universal mind might exist. I could feel it in the land. It felt neither malevolent nor benevolent, rather regally indifferent.

Captain Scott, hero of ultimate failure, believed in his men. He believed in luck. In recounting Scott's death in *I May Be Some Time*, Francis Spufford shifts into a fictional mode and assumes the consciousness of the explorer:

> *Sometimes you wake from a dream of guilt or horror that has filled your whole sleeping mind, a dream that feels final, as if it held a truth about you that you cannot hope to evade, and the kind day dislodges it bit by bit, showing you exits where you had thought there were none, reminding you of a world where you still move among choices. Day has always done this for you. It seems unfair that it should not, today.*

"Scott can make no effort that would change anything," Spufford writes. The Antarctic does not give you second chances, and perhaps no chance at all. As is well known, Scott was lax on planning; he believed a decent man deserved a decent chance. The lesson seems obvious: not to abandon ourselves to such an unstable and capricious force as luck.

"Great God!" Scott said, on arriving at the South Pole. "This is an awful place!" Amundsen, the victor, said, "Farewell, Pole, I fear we shall not meet again." Both likely spent only minutes at the pole. Then they turned around and started back. So much survival, and so little living.

At Shades of Light, I stare at the numbers I have written in the sales column. My eyes water. I am tired from working so many long days, from the pressure of the future that makes its presence known to me more and more as the days pass. In only six weeks, I will be on my way to a distant city, and university.

My eye is attracted by a glimmer. One of the crystal suns. A ray has pierced it, refracting the beam into four or five separate shards of light.

Then he is in front of me. I didn't even hear him come in.

There is a tangled wreck in his eyes. I am able to see right through him.

"Don't come here."

"Why not?"

"Because. I'm at work."

I look around hopelessly. The crystal sun has a moon which orbits it, kept at a certain distance by a thin pewter wire.

"Then meet me later, when you finish."

Later we walk along the river. Anyone passing us would think, *A father and a daughter — of course, they look like each other — out for an evening stroll.*

We stop by the edge of the path. He takes my face in his hands. I freeze. The look in his eyes is depthless. They are as flat as seeds. "I'm losing you."

"You never had me," I say.

"It's possible to lose something you never had," he says. "One day you'll understand."

Now they have a make of car. Michael tells Donna and me a car was caught driving too slowly near the old bridge. When the patrol car started up its blue light, the driver sped away. He managed to evade the policeman.

"If it was me driving, he wouldn't have gotten away." Michael shakes his head. "That's the exact spot we went trawling," Michael says. *Trawling* — the word hits me in the stomach. But then that is what I am, after all. Bait.

What kind of car could it be? Jeep Wrangler, K-car, Plymouth Sundance, Chevrolet Corvette, Honda Civic, Audi 9000, Ford Taurus, VW Cabriolet, Mazda RX-7. The car names are so familiar, so much a part of my life, but I will forget them, they will become as otherworldly as the numbered call signs of distant planets astronomers discover weekly. I don't know yet that I will manage to forget almost everything, not only of that summer but of the six years I spent in that town.

"Mazda RX-7, black. Blacked out windows," Michael reports.

Donna laughs and I join in. "Stoner car. Too obvious. It can't be him." We think the murderer will have a family car — a Ford Fairmont, say.

I remember the night before, walking on the highway shoulder, the empty crunch of gravel, car horns tearing through the wind as they sped by, then disappearing into the blue of distance.

A week later I am walking toward the apartment where he is staying when my suspicion is aroused. Suddenly everything has an etched quality; the casualness of the world has been erased. I see the drooping elms; two goth-type kids of junior high age, ice creams a caustic purple in their hands; cars sidling by, always, always obeying the speed limit; a fat woman walking a tiny white dog.

The purposefulness of what I see can augur no good. I dismiss it, because I have no explanation. It is not dissimilar to what will happen many years later on a ship in Cape Town, waiting to depart on a second trip to the Antarctic.

There is a fear, ready-made. *I may not live past this summer.*

I stop, try to catch my breath. My heart is pounding, my head feels like it will evaporate. I sit down on the curb, afraid I will faint. I raise my head and see heat waves convecting upwards from the asphalt.

This is what I have been waiting for, the signal from myself that I am in danger. I must stop my trawling missions with Michael. If I do it again, something will happen.

That night I am on shift at the Executive bar. There I will wear a white skirt, yellow top, white sneakers with those little ankle socks that have just been invented, inspired by tennis players. I will serve men — dentists and insurance salesmen — with stalled blue eyes, handing them plastic cocktail glasses amid splashes from the pool where their offspring play, the sun's lazy censor's eye overhead.

How different my father's eyes are from these blue-eyed men. Dark, restless. Sharpened, I imagine, with knowledge and the things he has seen from above: migrations of elephants through fat floodplains, caribou treks among pastel fields of wildflowers, mustard tundra. He is not lying. I can see their reflections in his eyes.

4. WINTERING

rotten ice
Old ice which has become honeycombed in the course of melting and which is in an advanced stage of disintegration.

MARCH 28TH

The ship has left the Falklands. Simon announces this over breakfast as another snowstorm blankets the runway. The ship has been delayed by an engineering problem; it is coming to pluck us out with only one of its two engines functioning. But the weather is bad: fifty-knot winds, the limit of what the ship can cope with. Normally the ship would not put to sea in such a condition, but there is no other way to get us out, and they are running out of time.

The news confirms my fears that something will go wrong — the ship will not make it across the Drake Passage on its single engine — that we will be left there.

That afternoon the weather cleared to leave us with an oyster sky. Shoals of ice had been shoved into the bay, where they were broken up by the gyre-like giant sheets of scrap metal. On the edge of the runway we saw a single fur seal, its back arched as if in supplication, staring up into the clouds against the tobacco light.

In the evening we saw nacreous clouds, also known as polar stratospheric clouds, which are common in winter in Antarctica. They appear as waves or funnelled clouds and shimmer with a raw mother-of-pearl effect. Their iridescence is beautiful, but they are also associated with low or zero ozone.

Our world was being emptied. Like a bar at closing time, one by one they went off into the night. Most of the Antarctic creatures present in summer — the albatross; skua; petrel; and shag; the crabeater, leopard, Ross, Weddell, elephant, and fur seals — had departed for warmer climes.

The announcement came over the bing-bong. "Orcas!" We all scrambled. Pods of whales had come right up to the end of the runway, bobbing their slick black heads above the water, staring us straight in the eyes, as if to bid us goodbye before setting sail for South Africa or California.

"They look like they're trying to tell us something," Mark said.

"They are," Simon replied. "They're saying, 'Get out of here while you can!'"

The winterers were eager for us to go, Glen the carpenter confirmed. "As soon as we leave, they'll make the place homely. Do a scrub out, get everything as they like it. Put the tables away, get some comfortable chairs in the dining room."

"Then what happens?"

"Then they settle down to watch *The Thing*. That's the ritual as soon as the ship goes. It's this film about an alien invasion of an Antarctic base in the winter when no one can save them. It's really good — the alien eats the dogs first, then everyone on base."

The winterers wanted only to lock themselves away from the world. Rob the field assistant told me, "To spend a winter in the Antarctic is something few people on the planet get to do. It's like being on another planet. They're keen to get going with their unique experience."

That afternoon, as another dank pre-night took hold, I had an obscure urge to read "The Rime of the Ancient Mariner" again. I hadn't looked at it since Caroline the diver and I had performed our ill-starred recitation on the ship.

I looked for it in the library but couldn't find it. In the end I looked it up on the net.

Reading the poem again, I was struck by the vividness of his lonely vision, the remorse, horror, fear, longing, anxiety, confusion.

How violent are the mariner's dreams, how terrifying the ice, his strange delusions of an abyss that is alive, beyond his fantasies of perfect maps, of science. The mariner is convinced that the icy realm into which he has wandered is more than it appears: it is a portal to the cold dread mysteries of life. The ice rises to fill the horizon of the poem, glittering with threat, much as Stephen Pyne describes the advance of the continent in *The Ice*: "The ice field rises and shreds into a mosaic of mountain and glacier, like a mixture of planets, all rock and ice, among stars."

And there, in this cold diorama, next to the corpses of the mariner's dead crew members, it is, hovering — Coleridge's white vision, as totally white as the strange visions I was having and which I became convinced were connected to the onset of winter, luminous and terrifying: an angel.

That night, to celebrate our departure, base held an End of the World Party. Three people came dressed as Death from Bergman's *The Seventh Seal*; one of the Marks came as the real thing — a Weddell seal with the number 7 emblazoned on his foam costume.

Jonah the gigantic engineer dressed as Nostradamus. He'd tried to dye a white bedsheet brown by tipping in a jar of cinnamon. He smelled like a walking hot cross bun. I came as a Mayan astrologer, dressed in a white tunic made from a bedsheet, but everyone thought I was a patient from a mental asylum.

We were all friendlier now that the ship was on the way and it had become unlikely that we would be locked up with each other for the next seven months. We shared impressions of our impending separation. The departure of the last ship of the season in Antarctica has a finality that echoes leavings in other, less connected ages — the ship pulling away from the coast of Scotland, laden with immigrants, headed to New Zealand, say, in 1790, or the polar expeditions themselves, seen off from the quay at Portsmouth or Southampton with fanfare and little certainty of when, if ever, the ship and her company would return.

Now, there are few places in the world you cannot buy a ticket out of or extract yourself from. Even those parts of the Arctic I would get to know in the years to come are accessible; northern Greenland is served by Air Greenland helicopters through the winter, and in Svalbard the SAS flight from Tromsø gets in and out most days.

On one of those final nights on base, I had a dream about our departure. We were lined up in the dark early winter morning, those leaving on the ship, and those staying on the quay. It was not clear in the dream where I was — on the ship or on the wharf. Or I was in both places, a split consciousness.

The moment of separation was coming closer. I realized I was on land. I could only watch as the ship drew its ropes up, slid the gangplank in to the aft deck, and its thrusters began to hum. As in nightmares I could not scream or speak, nor move a single limb. The *Ernest Shackleton* pulled away into the gloom. I watched it slide around the massive hexagonal iceberg in Ryder Bay we called the Pentagon. It curved the edge of the peninsula and slipped from sight, and our futures — those of us on the quay, those on the ship — diverged.

That evening we took a group walk around the point to watch the icebergs collide. This was a favourite evening entertainment as winter drew near, a kind of Antarctic demolition derby. When the powerful local gyre of Ryder Bay crashed them together, we heard a deafening grinding, then a sudden crack as the weaker of the two succumbed. Porticos and arches tumbled in a flurry of ice crystals and snow.

A single arch iceberg lay grounded in the bay with a broken buttress, trapped in the grandeur of its disintegration. The sunset shone through the giant hole in its middle. We stood in silent contemplation: Simon, Jonah, a quartet of Adélie penguins in their dinner jackets, and me.

Winter was hurtling toward us like a blunt slab of time. The antidepressants weren't working; my brain was still taut and I couldn't sleep. Taking them had made me feel worse, as if I had run out of hope — in experience, in life. In myself. I didn't possess the tenacity and the endurance of the men who gave their names to this landscape, and in some cases their lives. I didn't belong here. But by not belonging to the Antarctic, it felt as if I did not belong anywhere.

Reading *South* by Shackleton in my office, I stumbled across a startling passage. Up to that point, the book had been the chronicle of a great adventure gone badly wrong, then the struggle for survival. But in the chapter devoted to Shackleton's risky break for rescue, in which he and two companions sailed in a leaky dinghy across one thousand kilometres of the most treacherous seas in the world to reach South Georgia, he wrote of their conviction that someone — an extra presence, luck, fate, destiny — had accompanied them through their ordeal:

> When I look back at those days I have no doubt that
> Providence guided us, not only across those snow fields, but
> across the storm-white sea that separated Elephant Island
> from our landing place on South Georgia. I know that during
> the long and racking march of thirty-six hours over the
> unnamed mountains and glaciers of South Georgia it seemed
> to me often that we were four, not three. I said nothing to
> my companions on the point, but afterwards Worsley said to
> me, "Boss, I had a curious feeling on the march that there
> was another person with us." Crean confessed to the same idea.
> One feels "the dearth of human words, the roughness of mortal
> speech" in trying to describe things intangible, but a record

*of our journeys would be incomplete without a reference to a
subject very near to our hearts.*

If these men, so rational, so morally and physically tough,
believed in the possibility of fate, of God, or merely of a spiritual
chaperone, why did I find it so difficult to conscience?

An interest in future-telling is one of the last taboos operative
in our culture. It is considered avid, unhealthy, the terrain of
charlatans. The link between crystal gazing and polar exploration
might be the theology of crystal itself. Scryers look into crystals to
see the relationships between patterns and turbulence, between the
micro and the macro, between "lattice and distributed force." As
Eric Wilson writes,

> *Crystals, glaciers and the poles form invitations to the
> uncanny, preludes to gnosis. The crystal gazer transforms
> the frost on the window into strange attractions. The explorer
> converts the unmapped pole into his inmost column. The
> climber transmutes the moony ice into a revelation of the
> ground of being. In each case, the beholder experiences a
> return of some repressed energy that forces him to evaluate his
> habitual distinctions between familiar and unfamiliar.*

We all have unconscious *terrae incognitae* in our minds. Carl
Jung proposed that in those unknown lands the future might also
live, or an awareness of it, and that we might intuit the future
because on some level we have already been there.

I was afraid of my future. I realized I believed there was such
a thing as luck. I had a suspicion my life might actually be located
outside me, programmed somewhere in a vast database, more a
string of consequences than an experience owned and inhabited.
I had lacked the courage to adopt an entirely rationalist approach
to life. I had certainly lacked the more regular kind of courage

that might see you through an Antarctic winter.

In *I May Be Some Time*, Francis Spufford writes, "Do we not all wonder if we are brave enough to do what these men did?" He writes of Captain Scott's "deliberate decision to inhabit the impossible situation on one's own terms, rather than flailing uselessly against it." Is this not a definition of courage, even of heroism, however quixotic?

It is hard to underestimate the centrality of tragedy to the Antarctic continent. But this was now an old story, and I wondered how I was going to supersede it, as a writer. The French novelist and literary theorist Alain Robbe-Grillet examined tragedy in *Towards a New Novel*. The civilizing influences enacted upon the modern consciousness condition us to accept tragedy, perhaps to even expect it, he writes. "Tragedy may here be defined as an attempt to reclaim the distance that exists between man and things, and give it a new kind of value, so that in effect it becomes an ordeal where victory consists of being vanquished."

Robbe-Grillet had a revolutionary idea about the aesthetics of tragedy: "Wherever there is distance, separation, dichotomy, division, there is the possibility of feeling them as suffering, and then of elevating this suffering into a sublime necessity." He takes issue with the "tragified" universe that enlightenment culture has built and which has come to its apogee in the twenty-first century. He asks, might it be time to reject our own tragedy as a dismal and ridiculous fate?

MARCH 31ST

The afternoon of our last day on base we go for a last zip around the bay in the RIB. We go to the lagoon, then to Léonie Island. Our mood is buoyant now that our departure is imminent. Andy and I make coffee for everyone in the apple hut on Léonie Island and we all sit down and watch the fur seals chase each other. Two of them sidle up to us and look at our biscuits longingly with long-lashed eyes.

On our return the water is choked with translucent tumbling growlers which we must thread through carefully. On the larger floes a few remaining Weddell seals relax. They give us winningly moist looks as we pass in our orange-and-blue padded polyurethane outfits that make us look like *Fahrenheit 451* firemen.

We turn a corner to find an iceberg blocking our return to base — an arch iceberg, with a hole in the centre. The ice buttressing the arch on either side is nearly rubble, like old crushed stones of a castle left to ruin.

Andy cuts the motor and we drift closer. Out of the corner of my eye I see him inspecting the arch, and suddenly I know what he will do.

He turns us in a tight circle, revs the motor, and we drive straight through the arch. At any moment tonnes of ice could fall and kill us all. As we zip through I look up, half expecting to meet my maker in the form of bluemint toothpaste ice. But then we are through, all of us gripping each other's shoulders for support, grinning. Another reckless quest, another death cheated. We have been let through the portal, and are on our way home.

APRIL 1ST

April Fool's. I woke this morning at four thirty a.m. with my heart pounding, the thought poised in my mind: What if the ship doesn't come? The conversation with Ben last night runs through my head, Ben saying one of the things the Antarctic teaches you is to accept circumstances beyond your control. "If you have to sit in a tent while the weather clears, you sit in a tent. If the ship doesn't get here, the ship doesn't get here."

I can't convince my heart to stop pounding. In defeat I take half a pill of Temazepam, but that only gets me until eight a.m. I stay in bed until nine a.m., if only to try to make the day seem a little shorter. This is something I have done only twice before in my life, in episodes of deep depression.

The light is not encouraging. The sun is a black disc. As in a solar eclipse, there is too much radiance to it to call what it produces darkness, strictly speaking. The horizon is lit with a purple light, neither day or night.

In the afternoon we gathered around the computers outside the science labs to watch as the red blip on sailwx.info, the marine website that tracks ship's movements, approached Jenny Island. We needed to time our welcome: if we stayed outside for too long to watch it arrive, we would freeze. If we left it too late, we would miss its arrival.

An announcement went round on the bing-bong. "*Ernest Shackleton* sighted in Ryder Bay." We all waddled up the slope to the Cross. The sky was the colour of thunderstorms. The Pentagon iceberg was unmoved, eating up most of the entrance to the bay; it could have made our rescue impossible, had it been slightly larger.

Someone yelled, "Looks like our big red taxi has arrived."

It materialized out of a blue night, threading among the cathedrals of icebergs. We were silent, watching its approach. Andy the departing GA lay down in the snow on his back, clad in his padded boiler suit, his arms crossed over his chest, and gazed into the sky.

The ship arced round the Pentagon. Surges of emotion rippled through me: elation, gratitude, relief. Then, suddenly, nervousness. As if it were all a trick, a joke, and the ship would dematerialize just as abruptly as it had appeared.

The ship came alongside the wharf, the gangplank was down, and suddenly we were on board and into another world of fizzy water and Coca-Cola; avocado and lettuce — in Antarctic lingo, *softies*, *freshies*; other people, *strangers*; the return of the tick sheet; the red leather banquettes of the bar; framed maritime photographs: the *Shackleton* in the North Sea approaching an oil rig, the *Shackleton* crashing through the ice field.

Elliott the dog-musher was on board. He and his colleagues had been plucked out of Lockroy on the way down. I hadn't seen them since the day we put them ashore from the *JCR* in early December. We hugged, our faces lit with Antarctic euphoria: *We made it!*

In the bar it was warm. For the first time in so long I took off my three layers of fleeces. The anonymous terror evaporated. My nightmare was over.

I walk through the town in the early evening. I have told my mother I am going to Donna's. My friends are always concocting elaborate lies when they have drinking parties in the woods, or want to have a couples' night, parked in trucks. I never have to resort to such feints. My mother never questions where I am going.

I knock on his door. There is no answer. I decide to try opening it, and the screen door gives way. The kitchen countertop is clean. Knives stand ramrod out of a knife block. The clock says three thirty p.m. It is now seven. It had stopped when he'd left, perhaps.

I sit down in the chair and hold my head in my hands while the evening darkness leans into me.

I don't tell my mother, or rather I tell her that he left town, finally, but that we said goodbye.

"Good," she says, her voice hard and swift, like a book rapidly shut.

I have only four weeks more in that town. In early September I will be on a train to a city sixteen hours away. There I will attend university and start my real life. My future careens toward me, and in the meantime I distract myself by drinking with Donna and her friends. We drink a bottle of Jack Daniel's and go driving. I am at the wheel, accompanied by a boy I don't know well. I lose control of the truck, and we are both thrown into a tree.

"You should be dead," the nurses tell me, diplomatically, at the hospital. They are fed up of course —with battered women, murdered women, with drunken teenagers certain they are immortal.

Despite not wearing a seatbelt and sustaining extensive bruising, I have not broken a single bone. My companion is similarly unscathed, bar a bruise and a cut. My face smashed into the dashboard so hard it left an impression, a face-sized dent. "It's not possible," they say at the wrecking yard, when I go to identify the car. "It's not natural," the police say. The police are so stunned by my feat, they ring the local newspaper. In a week's time an article will appear on my escape, calling it "miraculous." Still, I look like a grey squirrel with my expanded cheeks and two soot eyes.

When I tell my mother, I hear her familiar casual, filigree laugh. "Like grandfather like granddaughter," she says, that musical note in her voice so like the dangling wind chimes Elin sells in Shades of Light.

The premonition I had that afternoon a few days before I went to see my father and found him gone without explanation, that I would die that summer, was not quite misplaced. *You should be dead* — this is the mantra of the police, the doctors, my friends. How delighted they all sound to know it could be true.

That is when I first learn to listen to myself carefully for that hum of alarm, for the sense of a distant hand nudging me, trying to communicate across a vast and gelid realm. I won't hear it again for many years.

As for the murderer and attacker of women in that town of backwoods capitalists and Loyalist mavens, he is never caught. After the tae kwon do woman in June, there were no more attacks. I see Michael and Paul in their patrol car, driving up and down woodlot roads, looking for a car with blacked-out windows. Or for a girl in jeans and a neon green halter top, trawling the side of the road for wild strawberries.

But before I leave at the end of that summer I have my first seizure of anxiety — this is how I will refer to them — since childhood, since the last year I lived with my grandmother and grandfather. I begin to believe — not to fear, or fancy, or imagine, but really believe — I am narrating my life from the dead, or beyond the grave. I believe I am alive, but am already in the next life. This life is the dream, and the dream I had about my father and the river is real.

My father let me drown, or drowned me, out of shame. Shame because he had wanted something from me. He did not kill me, but let me die. There is a difference.

In the dream, just like the two young women who fell victim to the murderer, my body was found, but not right away. I drifted downriver, into the stanchions of the bridge. My hair became tangled there, snagging like the long resinous river grass where the ducks hide their young.

Sometimes, at the end of my shift, at five in the morning, I restock the bar from the grumbling ice machines that stand sentinel on each floor of the Executive motel. I leave my cart and walk out the back door and go down to the shore. There I watch dawn soak the horizon, the early morning sorties of seagulls and kingfishers.

The river stands still. I will have to learn to impersonate it, to take its stillness into myself. This is what men expect women to be — a static dimension. For men, I think, whether it is this woman or that hardly matters, they are all the same fluid substance. They need to drink you, to move through you, but they are on their way somewhere else.

A heron stands in the sludgy water between the reeds up to its stick-knees. I remember the island, those years in the woods with my grandparents, the small animals, squirrels, skunks, raccoons, that looked at me with moist eyes from the perimeter of night. Also the more fierce and volatile animals who live just beyond the edge of darkness in the forest where Donna and I take refuge, their eyes boring a red hole in our backs as we walk to our drinking spot. Yes, they are definitely there, watching.

5. THE SYMPOSIUM

polynya

A stable ice-free water space in or at the boundary of fast ice.

November, a year and a half since I left the Antarctic. The earth is warming faster than ever. In the summer, the Arctic sea ice extent crashed, melting to a summer minimum nearly half the average of the 1960s and twenty-four percent below the previous minimum set in 2005. The melt opened a navigable route through the Northwest Passage, the sea through Canada's 36,000-island Arctic archipelago, for the first time in human history.

That year I attended many conferences, talks, seminars on climate change and the polar regions. One was an ice symposium at the British Library. The symposium blended cultural and scientific approaches to the Antarctic. One theme was about how ice cores, glaciers, and field stations could be thought of as archives, spaces of knowledge that inform how we imagine and shape our collective futures.

Another panel tackled the human rights of climate change: Inuit activists have been protesting that the right to be cold is a basic human right, one which is being threatened by global warming caused by pollution emanating from other parts of the world. The speakers also evoke the plights of the inhabitants of Pacific Island states and low-elevation Indian Ocean archipelagos. What are our responsibilities — moral and legal — to those people whose lives are being changed forever by our actions?

Eric Wolff is a glaciologist and climatologist. He does not work out of Base R, which is one reason why I had not met him, but rather drills from Dome C in east Antarctica.

Wolff is speaking on a panel which aims to put the science of climate change across to non-specialists. "We talk about climate change, but how do we know which climate is the optimal one?" he asked the audience, which was mostly made up of artists, geographers, historians, and journalists. "We think it is ours now. Or is the optimal climate the one of five years ago, or fifty, or two thousand? The fact is, climate is relative and changing. What we mean is the optimal climate for our way of life. And that way of life may no longer be tenable. That way of life may have to change."

Anthropogenic, *forcings*, *parameters*: we soaked in the cold bath of those rinsing, clarifying words, the language of modelling and climate change. "We don't know what the outcome will be," Wolff said from the podium. "But we are on that road already, with 350 parts per million CO_2 already in the atmosphere. Even if all carbon dioxide were turned off tomorrow, the planet would continue warming by at least 1.5 degrees."

Wolff went on to ask the tough questions, the ones that extended far beyond the remit of glaciology: do we suffer from a preprogrammed fatalism? Why are we persisting in our European enlightenment and romanticism frame, talking aesthetically more than scientifically, revelling in gamesmanship, in apocalypse visions, when the hard work of adaptation, of creating the political will to deal with warming, are absent? He noted that we hang on the notion of fulcrums — the turning point, the tipping point, the vanishing point — a moment in time usually identifiable only long after it happens. You only knew afterwards that this was a moment in time and space when there was a recognizable before and an after.

After Wolff's talk the audience, some three hundred people, filed out for coffee and tea in a sweeping foyer. I saw familiar faces from the Antarctic world. I met one of the visual artists, a painter, who had gone to the Antarctic the previous year. We compared notes. It turned out he had felt the same oscillations between euphoria and crushing depression. His moments of gloom there were not personal, or not as personal as other episodes of melancholy in his life, he told me. "It was climate change that freaked me out. Now it's all I can think about."

We stood together, the painter and I, clutching our teacups, sobered into silence.

We had just been told that within two hundred years, give or take a couple of decades, the planet may well become uninhabitable for all but a few pockets of human life. It's a bit like a gradual nuclear war, this slow apocalypse, but worse, in a way: radioactive material ceases to be harmful after its half-life of 490,000 years. But we are approaching the point of runaway warming. The planet will not cool in time for us humans and the species we share the planet with. There will be no half-life for us, no point in the distant future when the malign will regress into harmlessness. We will never be able to claw our way back to the equipoise between climate and civilization we have enjoyed for the best part of the last two millennia. Despite this imminent danger, our innumerable collective actions continue to add to the steepest peril our race has ever faced. We know we are doing this, but we cannot stop. We know we personally won't have to face the consequences.

A related problem, it occurred to me that November day in the British Library's cavernous auditorium, was one of imagination.

We do not entirely believe in the future. The past and present convince us with their reality because we have lived it, are living it. But through a cognitive slip, I think we are unconvinced that the future is real and will arrive, particularly the future beyond our lifetimes. There is another blind spot in human consciousness — we are hopeful beings. Prodding ourselves with fearful scenarios of the future to try to jolt us into action seems to have the opposite effect: we rebel against such negativity and lack of hope, and take refuge in stasis.

Eric Wolff, with his modest but meticulous approach, reminded me what I have learned to admire about science in these past few years: its single-minded thoroughness in a search for the truth, its need for argument and disagreement in order to arrive at that truth.

What we were also talking about in these august surroundings, the auditorium's plush seats, the lights dimmed, is the revenge of nature, a sort of divine justice at work. Our dread of ice has shifted from the Romantic vision of a frozen to a melt apocalypse. Ice will herald a very human fall from grace, through our wilful ignorance of natural rhythms and the dependency of society on nature. Like the humans and gods alike in Greek tragedies, we will be victims of our own arrogance.

"But we are also adaptable," Wolff said. "We are living in a solution-oriented time, rather than a superstitious, fatalistic Dark Age. We are tough-minded."

Wolff's optimism was gritty, hard-earned. I had the impression that he had taken a decision to view our future with some hope, because the alternative was unacceptable.

We clapped in a chastised, awed applause, and Wolff left the podium, taking with him the stern, metallic ring of ice language: *tarn, firn, flux.* These words seemed to me more than terms; they somehow convey a tangled forest knit of rules: mathematical facts as stark as life and death, but also the possibility of change, or at least of motion.

Wolff's talk reminded me, too, of the mysteries at the core of ice: How do the molecules loosen in the melting process? What is the nature of the bonds between the hydrogen and oxygen as ice breaks down? Why do ice crystals form with a small space around them, like a moat around a castle? Why are no two snowflakes alike? Is this really possible? Why, when it is so cold, does it feel hot to the hand?

"We hang on the physical fate of the ice," writes the scholar and poet Anne Carson, of Sophocles' poem, the one I read on the ship and which compared the experience of love to a lump of ice melting in warm hands. Sophocles writes of how love takes shape through a series of crises of the senses, how crisis calls for decision and action. We have unwittingly entered into an age of, as Eric Wilson writes in *The Spiritual History of Ice*, "wanton melt achieved through greed and waste." Welcome to our tragic ecology. We don't yet know if this will be a primal catastrophe, or another sidereal slip into the flux of evolution — just another in a cycle of death in the name of renewal, of new life.

We filed out of the auditorium into the early November night. Day one of the conference was over. It was mild enough that we delegates milled about in the British Library plaza for a while, as if reluctant to disperse. It was good to be among polar people again — anthropologists, cultural historians, artists, glaciologists, geochemists. It really was a fraternity. Here, a stranger was not quite a stranger, because you had both lived in the Antarctic, or the Arctic. I felt that reckless list again into euphoria, like a ship leaning over too far after encountering an unexpected wave.

But the British Library was closing for the evening, and this was London — everyone had somewhere to go, somewhere they had to be. As I walked to my bike, it struck me that exactly two years ago to the day I had boarded a plane to Madrid, then another to Santiago, then the long hopscotch flight to the Falklands. It was the beginning of something. I was going into the light.

I met the man I call Loki at a conference on hunger, two years before I went to the Antarctic. For years I worked in academic publishing and attended conferences on politics, economics, sociology. Loki was a professor of political science, British, but for years he had been teaching at an east-coast American university in a tidy, prosperous town where he lived with his wife and daughter.

We carried on the stilted, if friendly, conversation of two people seated randomly at a dinner table. He specialized in failed states. Most of his work was in Africa, in the Democratic Republic of Congo, Somalia — all the top failed states of our epoch. He was tall and lean. There was a rigid, daring quality to him: he looked like a man accustomed to living among dangers.

His was of the most moving faces I have ever seen: strong, not cruel, not fine featured, but with no coarse notes. Each expression was a fascinating interplay of angles and creases. He had a sudden, quasar-burst smile.

I was sawing my way through yet another institutional chicken breast and looked up to respond to something he had said, and saw he was looking at me. His eyes had softened into a look which was unmistakeable. My eye went to the ring on his fourth finger, as if to check it was still there.

For the rest of the conference I observed him closely. Black eyes, in his mid-forties. There was something static about him, a wax figure of a handsome man. This artificiality followed him through space, as if he had just burst onto a stage. In part it must have been the handsome man's awareness that he is being watched. His movements were lean and taut, as if he were keeping something under wraps.

That I knew him, knew his face, expressed itself as a low hum of contentment. It was beyond the buzz of attraction; it was as if he emanated an energy completely different to anyone I had ever met. I had tuned into a lost frequency.

If I stood near him I felt relieved, as if I had come back from exile, from a long way away, with no hope of ever seeing my family again, only to find them in him. All along, I had been living in a howling, echoing void, without realizing it. Suddenly, in his presence, I felt complete.

A thought surfaced from a great depth within me: Who are you?

After the conference I tried to go back to my life, to live normally. But every moment shimmered with threat, bloodlust, but also a dark love. I was under the spell of a base elation. Also something that could easily have been fear. I knew how necessary this man was to my existence. I knew that mine had been a life lived half asleep. Now I had woken.

Nights became a place of bleached, inevitable torments. I woke in soaked sheets. The same cold fire that had taken over my innards set up camp in my brain. I was either unable to think straight or I was thinking with the greatest clarity of my life.

At the same time, I felt as if I had had a million volts of electricity put through me. Every cell was altered. For eight months, I did not sleep through the night. At four a.m., I would be jolted awake by a panic, a black elation of a kind I would not feel again until those final days in the Antarctic.

I saw him intermittently. Each sighting of him ignited within me a physical hunger, ragged and dark. I had never felt lust before, or not at that pitch, and for a while didn't recognize it. I felt as if I were dying. I couldn't keep food down. But I could drink. I began to eat less and to drink more, to try to quell the crashing symphony of anxiety my body produced. At work I roamed the internet for pictures of him and when I found one I felt sick to my stomach. Some days I actually threw up.

I began to be ill — an illness I would later learn was a bad-luck sickness, something that pounced when you were low. I craved his presence. More than that, my alarm stemmed from feeling that my existence was linked to his, but that we had been separated. I had found him again quite by chance, and my survival once again depended on him. If I could not have him, I felt sure I would die.

This was hysteria of a kind, of course, and I recognized it as such. But as with the anxiety episodes I would later experience in the Antarctic I couldn't seem to do anything to quell it.

We only saw each other at conferences, at meetings. We were

correct and friendly. Professionally he was astute and quick-minded, and seemingly principled. We never schemed to see each other, although we emailed from time to time. Over two years we saw each other many times, and each of those meetings I felt pinned to a wall. Even the thought of him, or meeting someone who knew him, made me tremble.

We never did more than kiss. But I have never had a more powerful kiss, authoritative, but also tender. Prowling and sincere. It was more than a kiss — it had the density of sex.

What did Loki feel for me? He was only capable of a cold adoration. I had taken up residence on the outskirts of his vision. I lived far away, after all. He had a wife and a child. I knew how these things worked — marriages and families lasted, even when there was not very much love, through inertia and fear, the pressure of daily baths and food and cars and childcare arrangements.

I had no such glue to weld me to my life. I gave up my partner of six years immediately, or he gave up on me. I wasn't sure what was happening and I didn't much care. I wanted only to feel that fever gloss of love.

My illness intensified. I didn't know what it was at first. I felt I was fermenting with neglect. This desolate lust I felt had nothing to do with my conscious life or my intelligence. There was something deeper going on, something beneath the patterns of lived life, of daily existence. It was larger than me, what I felt shimmering between us, something beyond rational thought.

In *The Spiritual History of Ice*, Eric Wilson writes of the difference between white (cosmological) and black (daemonic) magic: "The 'black' magician is a supreme egoist. He is bent on transforming the world into a double of his wishes, on controlling the forces that threaten the persistence and power of his ego. Though full of *hubris*, he is consumed by fear and desire — terrified by forces that compromise his ego and desirous of destroying these forces."

I won't recount the details here, but I would discover that Loki was a kind of decoy, a dank pear, the uncanny double who discloses the shadows of the unconscious, the dark interpreter that Shelley, Goethe, Wordsworth dreamed of. He existed only to project back at me a negative vision of my capacities — negative not in its usual meaning, but as in a dark mirror, like the Claude glass landscape mirrors Victorian aesthetes carried, to better admire the landscape. Why are you pursuing me? His response said. But then why give me licence for pursuit? I protested. He needed me to enact upon him what he wanted most to do himself, then to sanction me for it.

In Norse myth, Loki is an ambiguous character, with a complex and shifting persona: on the one hand he is the Trickster — he fools you and in doing so forces you to confront your weaknesses. He has an animal nose for weakness; it interests him, in a predatory way. In this incarnation he is not necessarily dangerous, rather mischievous: a character driven by multiple instincts, someone sent to enliven, shake people up, keep them on their toes, as well as to deliver complex fates to their individual owners.

The other face of Loki is the Devil — it was Loki who tricked Balder, the golden son of Odin whom the gods decreed could not be killed except by mistletoe, the one substance thought too harmless to exclude from Balder's invulnerability. Loki who tricked Balder's blind brother, Hod, the god of darkness, into brushing Balder's chest with mistletoe, killing him instantly.

In Loki, both the real person in my life and the mythic figure he brought me to learn about, I understood for the first time that there really were dark energies in the world, and indeed within us. There are forces of light and darkness, forever locked in combat. As adults, we can't be innocent to this Manichean duality. The task was to believe that white magic — cosmological magic — of which ice would become a part, for me, was the stronger of the two.

In the year before I went to the Antarctic, my life felt like a conspiracy I had wandered upon but not generated. I might not even have a part to play in this weird script populated, like a Shakespearean play, by casual black jolts of fate, by winter hungers and a castle of ruin.

I began to suspect I was being lured to the Antarctic by an intuition that it would reveal to me an inner mythological drama I needed to understand. Stephen Traynor, a Jungian psychoanalyst, writes, "Unconscious content will seek expression in the world, all the more so when the subject is unable to come to terms with it in the inner world." I was becoming aware of the paltriness of the conscious mind, how tangential its plans and awarenesses are to our existence. The unconscious mind meanwhile is powered by dark energy, fuelled on motifs, tropes, myths — an alliance of symbols and powerful narratives common to many people across time and space. These seek expression in our consciously lived lives, although we are not necessarily aware of their needs.

It's a frightening concept, in some respects, but also alluring — the idea that there is a separate self inside us. Jung called it the shadow self. In some people's lives it grows in power until it threatens to flood the will of the conscious self.

To accept this idea is to agree that we live in a dark fable. At any point the dangerous quests of the unexpressed self can overwhelm our lives. The idea, which has its best interpretation in Jungian psychology, is that certain things we do, or people whom we take into our lives, are an attempt on the part of the unconscious to expiate a tension that has become so powerful that it demands overt expression. It is the inner narrative turned outward.

This was the source of the anxiety that overtook me in the Antarctic, I came to understand in that year of symposia. I felt I would become trapped there, as I had been trapped those winters of my childhood and adolescence. I did not avoid the Canadian winter for twenty years because of a lack of grit or a dislike of extreme cold, but because of the emotional properties winter held for me, what I'd experienced in my childhood and adolescence.

My anxiety only bloomed after friends such as Tom and Xavier had left, the planes had left, and winter arrived, with only one way out. I am normally good at being alone. But in the Antarctic we were exposed to others' scrutiny in a way that is no longer familiar to most of us, or certainly not those who live in cities. There was nowhere else to go, no distraction, no escape. But unlike a medieval society, we were not kin either. There was no protection. Some part of me began to be overwhelmed and demanded that it be taken out of that place, before further harm could be wreaked, before memories could be unpeeled.

The Antarctic was an attempt to resolve inner conflicts in my existence. I would not return to Canada to do it; it was safer to enact it far away, under an upside-down heaven, in a frozen foreign colony. How appropriate that I would go to that continent at the bottom of the planet, that place onto which we project our dark fantasies, as much as our utopias.

The symposium ended on that Sunday in silence, with a sequence of photographs shown by Rachel Weiss, an American art historian and cultural theorist. Panoramic photographs, bleak in their intensity, they showed an infinite ice field, snow that went on and on, until it disappeared into a convex horizon.

The night Tom and I flew back

from the Ellsworths via Ice Blue, about 100 kilometres from base Tom took the controls back from me, so that I could look out the window. He dropped us low to the ground, only two hundred metres from the deck. We skirted the mountains of Alexander Island to the west, flying over sliding oxbow rivers and turquoise glacial melt pools. Some had amoebic, formless shapes, others drew sharp calligraphy with sudden upright characters.

Then the plane whipped over a cold white river, an ice stream. In between were long blank periods of ice field, so devoid of features that my mind — through fatigue or hallucination — began to produce a prairie, and I saw wheat, trees, a floor of flowers; faint things becoming visible, then melting back into the whiteness. I thought I saw wolves loping over the ice, but this was only the light, grey and feral, and the shadow of the Twin Otter with the sun behind it.

I want to remember this. The sheets of sea ice beneath us, the cold gold light of the midnight sun on the snow, my pilot friend beside me, an exhausted vigilance in his eyes. But just as quickly as impressions formed, they slid from whatever had produced them — neurons, synapses — before I could capture them.

Then we were on the ground, buzzing to a stop amid flying gravel. But nothing was ever over that Antarctic summer because there was no night, and everything was day, and thrill, and promise.

I attended the British Library symposium in part because I was preparing for another expedition. It had taken two years to put together, but in early February I was to return to the Antarctic.

In the meantime I had been to the Arctic, to Svalbard, where I undertook a writer-in-residency at the art gallery in Longyearbyen. How alive the Arctic is, when compared to the Antarctic! I couldn't stop walking around town taking notes, bending down to snap photographs of the tiny snowdrops and the meltwater coursing down the road.

At 79° N there were flowers, reindeer, Arctic fox, Arctic hare, running water, birds, polar bears. The air had the familiar cold sting of the polar regions, but the sun was blistering — warmer than a similar latitude in Antarctica. I told my Norwegian friend, Silje, that at 79° S in the Antarctic, the latitude of Ice Blue, nothing grows. There are no animals, not even birds. It is clinically dead.

Silje and I stood in the middle of Longyearbyen's main street at two in the morning. Most people were asleep, cocooned in rooms with tin foil taped to the windows to banish the perpetual light. Sun streamed down from the vertical coal hills that surrounded the settlement. At last I stood under the ellipsis sun again.

During the day I interviewed people in town about climate change and tourism. More and more people were coming to the Arctic "to see it before it melts," and Svalbard is the most accessible place in the entire high Arctic, serviced by daily flights from Tromsø and Oslo.

In the never-ending evenings, Silje and I would take a steep walk up an abandoned coal mine, or have a midnight barbecue at a rustic cabin outside town. We had to carry a .308 rifle with us everywhere: 1,500 people live on the Svalbard archipelago, but so do five thousand polar bears. Before we were allowed to stay in the settlement we had to put in time on the shooting range and demonstrate that we could shoot a bear target accurately.

In the summers the archipelago is increasingly ice-free, especially on the warmer western coast, and famished polar bears come into town to forage for rubbish, a reindeer, or humans. Two years earlier two teenaged girls went snowmobiling on top of the western ridge that towers over the town. One was killed by a young female bear; the other threw herself over the cliff to save herself and broke nearly every bone in her body, but she survived.

The Arctic sea ice summer extent was at its lowest in recorded history. I had finished the novel I had been despatched to the Antarctic to write. It would be published the following year. In Svalbard that summer, I began to write another book about ice, not a novel, but an exploration of ice in its more metaphorical meanings as well as an account of my travels in the polar regions. I realized I needed more time on the continent. I began to feel a hunger to return — although whether to the real Antarctic, or to the idea of it, I was not certain.

Back in London on an afternoon in late August, I began to square myself up to this project. The country was flooded. A few days before, London had experienced a tropical deluge. At a party I talked to a friend, Maxine, about this book I was thinking of writing, with the working title of *Ice Diaries*.

Maxine had read Jenny Diski's memoir *Skating to Antarctica*. "What struck me about that book was how much of it was about death — dead places, her suicide attempts, her wanting to die," she said. "How is yours different?"

"Mine is about not wanting to die. Or wanting *not* to die."

I tried to explain to Maxine that for a while I had needed to live a comfortless life. It was the only way to repair myself. The Antarctic had given me a searing white salve. Now, two years later, it seemed the discomfort, boredom, the panic and claustrophobia I had felt had been a necessary test. I felt sure it wouldn't happen again. I was ready to return.

When I left the party it was late and the streets were slicked with the most recent bout of rain. That summer it would rain and rain. The El Niño effect, a stalling of the North Atlantic oscillation, an unusually positioned jet stream — weather presenters came on television every night, smiling gamely among the little cloud graphics that carpeted the map of the United Kingdom behind them, trying to explain.

My grandfather is driving home at four thirty in the morning. He has been at the Legion in town for hours, possibly for days, playing cards, drinking. But he is an excellent drunk driver, and he can get himself home. He lives in a tiny trailer now, on the land of a woman who had been the wife of a friend of his, and who is now a widow. These days it's this woman he terrorizes.

The Trans-Canada swings upward onto the long hill that is the spine of the island. On either side of the road are two dark slabs of spruce. No one lives here anymore; an abandoned farmhouse stands to the left, slowly being eaten by trees.

There is no moon, so he does not see the patch of black ice. He is travelling fast, gathering speed for the hill, driving with one hand. In his other hand is a cigarette, despite the lung cancer, throat cancer, cancer of the kidney and the pancreas that have failed to kill him.

Black ice is invisible unless a light or the moon reveals it. The car disappears over the side of the road. It is late, and no other drivers see him leave the highway. The way the car falls, it can not be seen from the road. It is not until the following day, when the widow raises the alarm, that they go looking for him. Still there is no sign. Maybe he stayed the night with friends, the police say to the widow, but she does not buy it.

The next day, a farmer out in his fields spots the car, half covered by snow. For once, he had been wearing his seatbelt. But his neck was broken. He died from this, or from hypothermia, or both; the coroner said he couldn't tell. At the wake, I hardly recognize him. His body is deflated and torqued into a strange position. He has shrunk to half his size.

After the funeral, my mother and I go for a drive. The island is on fire with the colours of late autumn: carmine, amber. Trees so red they are purple.

"He would haul her out of bed in the middle of the night, grabbing her by the hair." My mother demonstrates by taking a bunch of my own hair in her fist. "And he would make her cook dinner for him and his drunken buddies. This at three in the morning, remember, and her pregnant, or with a child to feed."

My mother looks off into the distance, into that peculiar denim blue of the island's inland sea. "He was evil, evil, evil."

She herself left home at fifteen to attend college in the city. "I never wanted to see him again," she says. "But all the time, I worried for her. What would become of her there, in that house, alone with him. Now that I was gone."

What indeed? By the time I went to live with them, only five days after I was born, it was still happening, over and over: the booze, the violence, the guns and the killing — of animals, those were the real deaths. But also the mock executions in the living room, the gun cocked.

"I would never — never, ever put someone in the situation I was in," my mother says.

"What do you mean?"

"Having to go through that. His . . . madness. His violence."

"Why did you put your child in that position then?"

She stares at me — an empty, flat stare. "Who?"

She really doesn't know who I am talking about.

I remember something Denise the visionary said on the phone, when I telephoned her to tell her that her augury of me going to the polar regions had become truth. "You imagined it first," she said. "You must have, for me to have seen it. Everything is made up of thought. Even after we die. This is the basis of the whole of existence."

But if everything is made of thought, then I do not exist, because my mother does not know who I am referring to, when I ask her the question. *Who?* I do not even have the density of an idea, for her.

When I tell her it was me, and point to my chest — as if that would clarify things. A small light switches on, somewhere deep inside her eyes. She shrugs. "I had no choice. I had no one else to turn to."

I am sure this is true. It dawns on me that I don't know my mother. She had never spoken about her violent and unpredictable upbringing with our shared father, my grandfather. She dealt with it differently. She might be unsentimental by temperament, while I have a romantic nature given to idealizing people, to a belief that love is the paramount force in life.

I left the town in the woods the day I turned eighteen. I had no birthday celebration; I didn't need it. I felt as if I had begun a new life.

I never heard from my father again, just as after a certain period after leaving the Antarctic I had no more news of Tom or Max.

When I was twenty-one I moved to Britain. In the twenty-five years that have passed I have returned to Canada a handful of times, but only once in the winter. In the meantime I have lived and worked for long periods in Brazil, Central America, South Africa, Kenya, Namibia — tropical or subtropical countries, far away from eastern Canada and its glacier-gouged perimeter. My only real experience of winter in this tranche of my life has been the Antarctic; even my four trips to the Arctic have all been in summer, when the sun shines twenty-four hours a day and reindeer and polar bears patrol the streets at three a.m.

Everything is made of thought. Denise told me this once

again, the last time I spoke with her. I picked up the phone one day to call to say hello and no sooner had I touched the receiver than the thought flurried in to land like a bird: Denise is dead. I said to this thought, No, she is only in her mid-fifties, why would she be dead? I dialled the number with dread. A message in a man's voice told me that she had died a month ago, and that if I wanted I could donate to a certain charity in her memory.

I found the obituary online; also two small articles. She had died after an illness. I had no idea she'd been sick.

I can still hear her voice: educated, caustic but not unkind, sympathetic yet mildly impatient. She was practical and level-headed, so it was hard to believe that she spoke to entities she largely refused to call spirits. Yet she did know many things that were beyond intuition.

Everything is made of thought. In the Antarctic, when I could no longer control my thoughts, I had never, even on those gunshot nights in the trailer, been so frightened. I had realized it was not the Antarctic itself but the thrill of collective endeavour I found there, as well as human warmth, that returned me to my original enchantment with cold. But once that warmth was gone, I found myself back in the place I had been born.

I have shuttled between frigid and burning climates, only grudgingly living in the in-between places. I have known heat in my life, and cold, but sometimes not much warmth. In extreme climates we are sharpened by our vulnerability. We could die of cold, we could die of thirst. But the truth is most people die of neglect.

A fiction writer might well believe this, that thought creates our reality. After all we make up entire books, we write them into being on the power of the projection of our thoughts. All writers, perhaps, insist on making too much of things, of seeing life as a metaphor to be employed in literature, rather than lived experience. We are prey to fanciful notions: for example, that the Antarctic is not really part of the world, that it is more akin to living in the deep ocean, or on the moon.

But I know it is true: the Antarctic is completely unlike anywhere else on earth. In leaving civilized life behind and willingly entering into the Antarctic's charisma, exposed to its lethality, we can know who we are. What it means to be human. Whether we are human. Whether we can survive on a planet we are destroying. What we are destroying.

It is also a landscape, a physical terrain in the world. I should probably stop projecting onto it. I see the Antarctic for what it is now: somewhere separate from my consciousness and my origins. A cold place.

6. FLARES

glimmer ice

Newly formed ice within cracks or holes of older ice.

Cape Town in February. Two years after I left Base R, I arrived in the southernmost city in Africa to join another Antarctic expedition. It took me nearly a year of applications and presentations to be allowed to accompany a multinational ship-based scientific project. My project this time was to write the ice book I'd conceived in Svalbard.

We boarded the ship on a Tuesday; that night we were scheduled to set sail. But the ship, the *Southern Cross*, was delayed by high winds. On the Cape in summer the Southeaster blows — a powerful seaborne wind. The day we were supposed to depart it was gusting to forty-five knots and the container vessel carrying the scientific instruments could not come alongside the docks to be unloaded. In that much wind, the cranes that winch cargo from the ship to shore are unreliable, and the last thing you want is a container swinging wildly in the wind.

The chief scientist told me that delays aboard the *Southern Cross* were rare. Every day spent in port meant a day of lost science. "You can imagine, my head is full of red crosses right now — stations we'll have to skip. Moorings we won't have time to recover."

Once we were on our way, there would be no way off the ship for ten weeks. *Southern Cross* cruises are long because of the steaming distance from Cape Town to the ice shelf base — about ten days — and because of the multidisciplinary nature of the science done on board. On the way back, the ship would call at King George Island on the Antarctic Peninsula, then make its way up and across the Drake Passage on roughly the same route we took down to the Antarctic on the *JCR*. We would end up in Punta Arenas in mid-April. From there it was a short hop across to the Falklands, where I had arranged to do six weeks of scholarship-funded work.

As we waited in harbour for the weather to change, I talked to one of the helicopter pilots, Marcus, and a Dutch engineer, Pieter. I met my cabinmate Anneliese, a young Frenchwoman, and other young researchers. Everyone was friendly and excited about the cruise. The *Southern Cross* is one of the world's foremost polar oceanographic research ships, a well-appointed science platform, efficiently and safely run by the organization that owns it. A trip aboard it is manna in the polar oceanographic research community, and everyone was suffused with that heady Antarctic combination of thrill, enthusiasm, and conviction.

Marcus the helicopter pilot and I decided to go shopping at the V&A Waterfront. At thirty-seven, Marcus was one of the most experienced helicopter pilots in the Antarctic. He had just done a stint in Saudi Arabia flying for an oil company. He was well-spoken, gentle, but also military in bearing with aviator shades. His face had that sandblasted quality of white men who have spent a long time under the desert sun.

We talked about Base R, which he knew well. "Such a beautiful place," he said. "When I was there I took a walk around that peninsula, what do you call it? The point. I couldn't believe the facilities there. It's so much better than being aboard the ship."

"Why's that?"

"I like the ship, but only when I'm working. There's too much downtime. At Base R, you have so many activities: you can go skiing, you can go for a walk. Here, we are stuck."

The Southeaster blew and blew. The meteorologist on the ship told us we would not get a respite until Saturday.

It was on Thursday I first felt it. As when I was on Base R, it did not build, but pounced at intervals. In the nearly two years since

I had returned from Base R I'd had no episodes of anxiety. Those months had been ones of calm productivity; I had written much of the novel I'd been sent to the Antarctic to write.

But on that Thursday and Friday, I became aware that what I was feeling was not anxiety, or not exactly. It had in common the Dash 7 takeoff run speed of the attacks I had felt at Base R, but it was composed of a different substance, thick and viscous. After a while I recognized it: dread. Dread is distinct from fear in that it has an object, a sense of knowing what is coming. It felt heavier, more metallic than fear. I knew somehow that my dread was linked to an actual occurrence, not something I was imagining. It was as if I could feel the outlines of a solid block, like a building or a house, being built, somewhere beyond the perimeter of visibility.

I ran through possible disasters in my mind. The ship was even larger than the *JCR* and had weathered many Southern Ocean storms. Surely it wasn't to do with the ship. Each time I dismissed my fears, they returned, stronger. There was a dimension to them I had never experienced before: behind them was a voice, or a force. It came from very far away and pressed against me, like a plank. It said, Get off this ship.

I told myself to wait, that it would pass. I wanted so much to stay. With my rational brain, I mustered all my arguments about professionalism and responsibility. Yet as I walked around the ship I could not escape a sense of unhappiness, even despair. I had a vision of myself in the cabin I shared with Anneliese, anxious and frightened.

The news came that the wind would abate at last, and we would sail on Sunday morning.

All Friday night I could not sleep. My heart pounded and my brain shrieked. I had never spent such an awful night, even during my lowest moments at Base R.

In the morning, I got up and without thinking went straight to the bridge. There I found the captain and the chief scientist and made an excuse about a family situation. I left the ship the day before she sailed. It was a mess: immigration formalities, excuses, awkwardness with shipmates, people I might have called friends.

By the end of the day, I stood alone in a sun-bleached parking lot outside the Cape Town customs house, surrounded by Zimbabweans and Malawians trying to regularize their immigration status.

I remained in Cape Town for two months, too paralyzed by guilt and frustrated to return home. These people had given me a much-desired place aboard their ship, and I had let them down. Let myself down.

I could not tell anyone of the real reason for my decision, that I thought something would go wrong with the ship, or the expedition. I had no proof and I would only create unease. They would think me crazy, and I wouldn't blame them. *I* thought I was crazy.

That last morning on the ship when I was talking to the chief scientist, my sleeplessness and frustration got the better of me. I began to cry. I had explained I would have to leave but was unhappy about the circumstances.

"It is a pity," he said, "but don't worry. What can you do? We all have to live with ourselves."

I returned to London, from where, in late March, I would take the flight from RAF Brize Norton to the Falklands, to begin my fellowship in the Islands. I was supposed to travel to the Islands from Punta Arenas, after leaving the *Southern Cross*. But I was now in the wrong place, and the only way to correct it was to undertake a carbon-costly journey to the Islands via the UK. Just after dawn on a late March morning, I would find myself on Ascension Island again, cooling my heels in the Cage between legs of the RAF flight to the Falklands.

The email arrived from a friend in the Antarctic world, one of the few people who knew I had left the ship. It began, *Did you hear?*

Less than three weeks after the *Southern Cross* finally left Cape Town, a helicopter piloted by Marcus and carrying, amongst others, my cabinmate Anneliese crashed on the Antarctic ice shelf. Marcus and Pieter the engineer were killed. Three others, including Anneliese, were seriously injured.

I don't know if I would have been aboard the helicopter, but there is a good chance I would have been. The organization took writers seriously. As the ship's writer in residence, I would have been at least encouraged to go and see the terrestrial base. I wouldn't have had any reason to turn this down; I liked and trusted Marcus immediately. He was a good pilot; my time with Tom had taught me that you can just tell.

The ship carried on, after the two bodies and three injured people had been evacuated by air. No one else was allowed to leave the ship. For seven weeks after the accident, they were at sea. One helicopter and one pilot remained to do the ocean mooring retrieval work.

For a long time the only thing I could think about was Marcus, our quick complicity as we stood on the deck of the ship after coming back from our shopping trip, the last chance to stock up on consumer goods for many weeks to come. "It was so beautiful," he said, not of Cape Town and Table Mountain, which shimmered in the heat in front of us, but of Base R. "The mountains, the icebergs. I wanted to stay there forever."

APRIL 4TH

Flares zip open a dark morning. The sky to the north is lightening slowly, a blue oblique light. The cold throb of The Ice is loudest, just as the dark is darkest, just before dawn.

We are on the *Ernest Shackleton*, leaving Base R. Although I have not sailed on this ship before, it feels like a homecoming to be back in ship life — the tick sheet, the gash rota, the survival-at-sea drills.

As we curve out of Ryder Bay, there is no sea ice, only the berg, the massive girds of the Pentagon. The *Shackleton*'s captain confirms our suspicion that if the berg had become grounded toward the east, it would have blocked the ship's approach. "It's the biggest one I've seen close to base for years," he said. "We're lucky, this time."

The voyage out on the *Shackleton* has almost nothing in common with our journey down four months before. There is none of the shock of sailing right through the veil that separates the real world from the Antarctic. The surprise and euphoria are gone. We are subdued, chastened. Something is definitively over, and now all we need to do is get home.

We inched up the peninsula. The first two days at sea we encountered persistent blizzards. The seas built until we were ploughing through seven-metre swells and shipping spray over the fo'c'sle every three or four wave cycles. The *Shackleton*'s signature corkscrewing motion began. Many people get seasick on the *Shack*. It pitches normally, but its empty aft deck, a requisite for its other life as an oil rig supply vessel in the northern hemisphere summer, means that it does a semicircle turn at the back in the wave troughs. It feels like being on a giant pepper grinder.

I joined the second officer on the bridge as the ship pitched and yawed through increasingly confused seas, snowflakes driven against the window fast, like quarks, fireworks, subatomic matter harried to the speed of light. They glowed beautifully, illuminated by the ship's searchlights.

We were the only thing alive for many miles, apart from the *Southern Cross*. We had passed each other a few miles back, in the night. I saw the ship only as a garland of light laid across a dark horizon. The next time I would see her would be at the dock in Cape Town, two years in the future.

I went back down to the cabin I shared with Melissa the doctor and tried to get to sleep. At one thirty in the morning, I woke to find myself standing up in my bunk. The ship righted itself enough for me to clamber down to the listing floor. Spooked by the angle of the ship's roll, I dressed quietly in the dark and went up onto the bridge. I found the second officer at the helm.

"What happened?"

"The ship just did a maximum roll," the second officer's face was grim. He pointed to the gyro, the mechanism that stabilizes a ship's roll. "A wave broadsided us on the beam." The average wave height that night was ten metres with forty-knot winds — not  particularly bad conditions for the Drake Passage. As long as the ship's engines kept running, and the ship was headed into the wind and the wave roll, the conditions didn't pose any danger to a ship the size of the *Shackleton*. But this wave had been twice the usual size and came out of nowhere to broadside the ship. I'd woken standing when the ship hit the trough.

The following day the storm abated — enough that the ship was mainly pitching, a far less alarming motion than rolling. In the lounge everyone was watching films — *The Bourne Supremacy*, *Enemy at the Gates*, *Two Weeks Notice*. I went up on the bridge at sunset. I watched snow petrels and black-browed albatrosses skim the waves, which were blue and orange in the reflected sun. Cloud coated the horizon. The sunset moved within it, a smudge of peach.

The beauty of the Southern Ocean and the Antarctic returned to me, now that I was going home. In my last weeks at Base R, I couldn't see it. Anxiety coagulated my thoughts, made them an indivisible substance, a burden to carry around, a lump. Now I had rejoined myself and I could think clearly again.

A line came to me from a book I had read on the Antarctic although I couldn't remember which — *I felt as if I had come back from another planet.* My instincts were right, in a way: I had to be there until the end. The moment of leaving the Antarctic was like the moment that divides life from death, even more so if you are convinced, as I had become, that you would die if you stayed there.

Just as in the dream I'd had, when we left Base R that final morning we all formed two lines: on the ship's deck, the leavers, and on the wharf, the stayers. The winterers lit expired emergency flares. The plume they made as they seared the red Antarctic dawn was chemical and livid. It lit the black water in front of the wharf as if it were on fire.

We looked down from the deck to the winterers. I saw Caroline the diver with her apple cheeks and uncombed hair, having fallen out of bed to bid us our early morning goodbye, dressed only in her fleece.

The ship's thrusters nudged us away from the wharf. Slowly, we watched the gap grow. One, two, three metres of black water lit by fire. Then the ship turned and pirouetted on its haunches. We steamed out of the bay, pointed toward the far side of Jenny Island. Very quickly base receded. We remained on the outer decks for a while, as did the winterers on the wharf, despite the cold and the biting wind, until we couldn't see each other anymore.

It is minus thirty. I walk alone on a road that curves round to reveal a farmhouse in the middle of a clearing. The moon is silver and black, a celluloid negative. The sky is clotted with stars.

For six weeks between mid-December and early February, the temperature remains between minus fifteen and minus thirty-five. At the lower end of this scale your flesh can easily freeze, especially if there is a wind chill.

I am at stables where I work taking care of horses in exchange for riding lessons. It is thirteen kilometres from our house, out in the middle of nowhere. One night there is no one to give me a lift back into town and I can't stay at the barn — overnight temperatures will be minus twenty-five.

I ring my mother and ask for an emergency lift, but she refuses.

"I'm going to have to walk," I say.

"Then walk."

It takes me over two hours to walk home. I twitch my face all the way, in order to keep my cheeks from freezing.

It is a quiet night. Overhead the stars look as if they have been cast from ice crystals. Pine trees line the sides of the road in crisp silhouettes. The houses I pass have a look of sanctuary, like houses on advent calendars, the buttery squares of light in their kitchen windows, a red glow of togetherness just beyond the frame.

Walking along the road that night, I begin to construct, for the first time, dreams of the future.

Love is abstract to me, until attached to a person. An image is forming as I walk, one eye on the road as the other casts for a possible escape route thorugh the garrison of trees should a pickup of drunks appear.

I see a coal-eyed man, his face is handsome. He is thin but muscular. What strikes you is the intelligence in his face, how fluid it is, how easily it changes from one expression to another. But there is another element that can't be named — an amalgam of instability and ruthlessness, perhaps.

It might be that people construct their lovers before they meet them, as much from their deprivations as from their desires. He is real, this man. He glitters like frosted kindling. Yes, a pale, dark-haired man; he has done something unusual and daring in the world, and it has brought him closer to a suite of eternal understandings I want for myself. I will want his experience, his knowledge, as much as I will want his body.

The walk that night is cold and long but very beautiful. I still remember how, as I rounded the curve in the road beyond which the town's lights would become visible at last, I heard the call of an owl ripple through the frozen forest. I arrive at the house I share with my mother and her family hungry and cold but enlivened by this small feat of survival.

EPILOGUE
BLUEFIELDS

June 21, "the still point of the turning world." Latitude 51.52° N, longitude 0.10° W — London.

Today is the longest day of the year; we think of it as the beginning of summer, but technically it is the beginning of winter in the northern hemisphere. The sun rose today at four forty-two a.m. and will set at nine twenty-one p.m. Tomorrow there will be one minute less of daylight: the sun will rise at four forty-three and then set at nine twenty-one. For the following three days, June 23, 24, and 25, the sun will rise and set at exactly these times, while the planet stalls, before tipping into winter.

For now, though, the light is long and blue. Trees and the spires of churches are cast in iron silhouette against the sky. On clear nights a bronze glow persists on the horizon well past eleven p.m.

In the Antarctic this day marks the beginning of three days' midwinter holiday, complete with Christmas trees, winter sports games, and a midwinter gift exchange. The chef prepares a turkey feast, and there are to be days of drinking, a ceilidh. At Base R the only light is a dim glimmer in the sky, visible from eleven a.m. until one p.m., to the southwest. But at that latitude change comes quickly; on July 22, less than a month from now, the sun will once again peer over the horizon.

I stare into my computer screen in a darkening room lit by the lingering light of summer. The news trills with reports of record thaws in the Arctic. Also food shortages, artificial intelligence, surveillance, the credit crunch: our future seems to be crystallizing. But the future is here, I realize, just unevenly distributed.

Some effects of climate change — a perceivable average warming of the world's oceans, increasingly violent storms — are making themselves felt much sooner than even the most pessimistic of climate models predicted. The most radical effects are in the

Arctic. Here the melt season has lengthened by more than a month since 1979. The Greenland ice cap is losing more than 200 gigatons a year. Apart from the obvious effects of coastal erosion, habitat destruction, and species extinction — the polar bear in particular — this seismic shift in the morphology of the ice cap will have many ramifications, reaching beyond the far north.

All that water pouring out of the Arctic ice cap needs somewhere to go. It flows into the oceans, increasing their volume. Coastal erosion is already beginning to make itself felt in the UK, as we could see in the damage caused by the fierce storms in the winter of 2013–2014.

In the meantime, melting polar ice has opened a northern sea route to shipping, both commercial and military. At least one-fourth of the world's undiscovered oil and gas resources is said to be located in the largely frozen Arctic Ocean. Nearly eighty-four percent of the ninety billion barrels of oil are located offshore. In these last-gasp decades of dwindling accessible petroleum sources, it's unlikely these resources will go unexploited. The retreat of Arctic ice and the sea routes now navigable for the first time since seagoing ships have existed will increase the risk of competition for energy resources.

Will we witness the dawn of an age of thermopolitics, a new Cold War? Wikileaks cables published in 2011 quote US diplomats referring to "the potential of increased military threats in the Arctic." At the same time the Russian ambassador to NATO was quoted as saying, "The twenty-first century will see a fight for resources, and Russia should not be defeated in this fight . . . NATO has sense where the wind comes from. It comes from the North."

A newly aggressive Russia has designs on this Arctic motherlode. Putin believes that the Arctic is an essential ingredient in Russia's ability to maintain its position as one of the world's largest oil producers. Russia gets fifty-two percent of its budget revenues from oil and gas, as well as seventy percent of its export earnings.

In 2012 Russia produced an estimated 10.4 million barrels of oil per day; nearly two-thirds of that came from western Siberia. But many of Russia's oil fields are starting to decline. The Arctic is a logical place for Putin to expand.

In Antarctica, off-limits to resource exploration until the expiry of the Antarctic Treaty in 2048, the picture is more complex. The East Antarctic Ice Sheet continues to accumulate mass while the western Antarctic, where Base R is located, behaves much like Greenland, draining ice into the ocean. Each year the ice streams, such as the one Oddvar and his team radared, siphon ice from the continent to the ocean at an accelerated rate. Glaciologists estimate that two-thirds of the continent will accumulate mass through climate-change-generated increased precipitation, while the other third will lose it through glaciers racing down slopes, shunting their cargo of ice into the sea.

The sea ice picture is more complicated yet, with ice growing in some areas and shrinking in others. Antarctic sea ice dynamics made the headlines in late December 2013 when the Russian research vessel *Akademik Shokalskiy* became stuck in thick sea ice off the coast of Antarctica, in an extended version of our entrapment on the *JCR*. After an unsuccessful rescue attempt by other research ships, its fifty-two passengers were eventually evacuated.

All this information blares in my mind like a television picture hazed with static. It seems to have happened so quickly — the change in our circumstances, the change in the earth's climate. And we are trying to respond, equally deftly, to manufacture the requisite alarm to do something: witness the films, the documentaries, the computer-modelled scenarios of our future.

I believe the mind can scan the future, that we have our own computer-modelling software encoded in our brains, but we don't know how to use it. I am only an averagely prescient person, but through having an anxious nature I have sometimes become more alert to the possibilities lining up on the wave tips of time.

Occasionally our intuition transmits what the mind sees there to our psyches.

But largely we are protected from knowing too much about the future. Unless, of course, you indulge in augury — clairvoyants, astrologers, cards. To know the future exposes the human necessity of living with the aid of the helpful illusions we call desire and hope. Very often, the truth of the future erases these consolations from our mental landscape, leaving us undefended against its certainties. The truth is a cold and blasted place. Are we sure we want to go there?

I suspect that the future exists in that timeless and spaceless realm inside us, just as does the past, or the present, and we are generating it, at least in part. As proof of this I offer that the future can be predicted — to an extent, at least. The difference between the future and the past is that the future is still subject to change. Although how much change, we don't know. We might live in a garden of forking paths, but all paths might lead to the same destination.

There is another possibility: that we are dreaming our future into being. This is not only an individual dream but becomes, through a synergy we are not able to understand, a collective one. This collective dreaming — what Jung called the collective unconscious — is manifested imaginatively. I worry that with every thrill-inspiring apocalypse film about flooding, tsunamis, giant storms, we are shaping ourselves to a particular future, one which is only a projection of fears.

You can make what you fear happen simply by focusing on it. This is a cherished idea of New Age thinkers, but it has been around for a long time. It is what anthropologists call magical thinking, the belief that thoughts have agency, that how you think about an event can affect its outcome.

It was Xavier who reminded me what Ernest Shackleton said, after the crushing debacle of the *Endurance* expedition:

"A man must shape himself to a new mark directly the old one goes to ground." The secret of Shackleton's flexibility, and his ultimate survival, might be that he was always looking for the next challenge, and he was able to quickly put disappointments behind him.

In Shackleton's *conseil* there is a message, for all of us, in dealing with the conundrum of ice and its loss, and of global warming. We must shape ourselves to a new mark, to carbon emission cutting. But even with cuts, our planet will go on warming, and we must adjust. This will test the limits of our ingenuity, our tolerance and compassion, our flexibility.

But meanwhile we seem sufficiently besotted with apocalypse that we are projecting it into the future. There is a reason why so many apocalypse-mongers are religious: at heart there is a spiritual yearning to be done with it, for the world to be razed, so that new beginnings can take hold, but these new beginnings are usually of their design. Religious people have a longing for release, for the single event which will encompass and explain everything, like an explosive matrix which, in a flash, reveals its architecture.

Subconsciously, all civilizations seek decimation. We are aware of the caustic design in the machine, the bad debts, the ruthlessness, the ends-justify-means-ness of human life. We want to be absolved, cleansed, and so to begin anew. Every apocalypse advocate assumes that in the painful nightmarish sorting of the wheat from the chaff, they will somehow be preserved as kernels for the future. Human beings have the genetic arrogance to manufacture this delusion. Fed on diatribes of obligatory heroism, we don't imagine it will be us who will fall at the first hurdle.

You can feel it, as you approach. Its presence is like that of nowhere else on earth. The monumental self-absorption of the landmass acts like a cold vortex, pulling you in. But the allure of its independence, its lack of need, is so attractive. The Antarctic lives outside our narrative, like an extra planet moored at the bottom of the ocean. It does not belong to us.

Sea ice is the most accessible of the ices: you can watch it drift by, crystal and sure, an ice ship at sail in these underworld waters with its scars — the stigmata of stamukha, rivered with sastrugi. From its windscoop peaks, seracs hang in suspended flux. In other places, raked by katabatic winds, firn is exposed, its metal gleam ablated. Heat competes with a deep, monumental cold. Soon the albedo cannot resist the sun's broiling gaze, and melt begins. The process is not gentle, as a stream might melt in the spring, gurgling and trickling, but a frozen volcano erupting from within. The shear and stress of melt is that of any transformation, achieved only through a spasm.

That Monday morning in early April when the *Ernest Shackleton* headed straight out into the wild, open seas of the western Antarctic Peninsula, is my final moment of departure from the Antarctic, despite my attempts to give myself another chance, a second experience in the continent that holds its acolytes rapt for years afterwards, mesmerized by a shimmering inner horizon they did not know existed before they went there. We were in a rush to get to the Falklands, and there would be no scenic route home. Sometimes paths are not meant to be re-taken. On the *Shackleton* that morning, I didn't know yet how in less than two years' time I would try to return, but that fate, knowing me done with this place, would repel my attempt. I will never come this way again.

For years I had Antarctic dreams. In one, I fly to a remote colony with clapboard houses and spruce trees where we drive 1980s cars down battered streets much like those of that town I lived in during high school. The Antarctic has become a proper country with a settlement, and this hardscrabble town will be my home but also a prison; there is only one flight a year, in the dream, a Continental flight via Houston.

In another, the Twin Otters and the Dash 7 are stored in an iceberg, not in an aircraft hangar. There is a secret door to the iceberg, a code you must know to open; I press it and find them again, gleaming under the hangar lights like reposing dragons. In another, icebergs menace the Falkland Islands far more effectively than the Argentine government ever could; they crash ashore, submerging the islands in a tsunami of instant melt.

Eventually the Antarctic released its grip on my subconscious. The dreams have stopped now, but for one.

Tom is flying me to Bluefields on the Ronne Ice Shelf. "I don't know who was possessed to call it that," he says, his mouth curving into his characteristic disapproving tilt. I know what he means — despite its lovely name, like so many "places" in the Antarctic, Bluefields is no more than coordinates on a GPS, just a few fuel drums marked by rickety bamboo flags, their red triangles ripped open by katabatics. There is no blue, there will be no fields.

We approach the shelf, banking in the Twin Otter over the gauzy muslin of frazil ice on the surface of the Weddell Sea. We land, skiing to a graceful stop on an ice sheet with the consistency and reflective power of chrome.

We get out and stand beside our little carmine plane, scanning the horizon. Sea ice. Loose pack. Ice pans. Glare ice. A steep descent toward open water.

We stare into the horizon and its barking, infinite light. The convection heat rays of the sun beat down with such intensity the landscape sashays. We stare at the white snowfields, fields of such deep, pulsating blankness they pass through an invisible colour barrier to become cerulean, indigo, tourmaline, turquoise.

After a while, Tom drags his gaze away, and we blink into each other's eyes. In the dream we are still friends. We look at each other, astonished. We say it in tandem: *Bluefields*.

SHORT GLOSSARY OF ANTARCTIC TERMS

albedo: the extent to which an object — here, ice — diffusely reflects light from the sun.

avtur: aviation turbine fuel. A kind of paraffin which has been specially treated to withstand low temperatures.

bergschrund: a crevasse that forms where a moving glacier ice separates from the ice above. An obstacle for mountaineers.

bing-bong: the popular name for the Stentofon PA system used on base.

col: the pass in between two mountains, from the Old French, *col* — neck.

CTD: in oceanography, an instrument used to measure conductivity, temperature and depth of the water column.

field: away from base. Working "in the field" means living in a tent unless at a big field camp. "Deep field" means living and working a long way from base.

field assistants: mountaineers charged with helping to look after scientists in the field and on base.

firn: crystalline or granular snow, especially on the upper part of a glacier, where it has not yet been compressed into ice.

frazil ice: cinder-like accumulations of ice.

gash: general cleaning and housekeeping duties of the day on Antarctic ships and bases. A Navy term.

HF/VHF: high-frequency radio.

hummock: a mound or hillock of pressure ice.

iceblink: a white light seen on the horizon, especially on the underside of low clouds, due to reflection from a field of ice.

Iridium phone: satellite phone.

JCB: a construction vehicle such as excavators and backhoes, made by J.C. Bamford Excavators Limited in the UK, universally known as JCB.

katabatic wind: a wind that carries high-density air from a higher elevation down a slope under the force of gravity. A katabatic wind originates from the cooling by radiation of air atop a plateau, a mountain, glacier, or even a hill. Most commonly found in Antarctica and Greenland.

manfood: as distinct, originally, from dog food. Manfood boxes are wooden boxes that contain ten days' supply of dried and tinned food for two people.

melon hut: an oblong-shaped fibreglass hut used for field operations.

met: the meteorologist; weather forecasters seconded by the UK Met Office to base.

nunatak: an isolated hill or mountain of bare rock rising above the surrounding ice sheet.

parhelia: bright spots in the sky appearing on either side of the sun, formed by refraction of sunlight through ice crystals high in the atmosphere.

p-bag: personal bag containing a sleeping bag, sleeping mat, and sheepskin, for use in the field.

pit room: accommodation; bunk-bed-style dormitory rooms on a (UK) Antarctic base.

PNR: Point of No Return. The point at which there is no longer sufficient fuel for the aircraft to return to its point of origin; PNR means the airplane is committed to land at its destination.

RIB: Rigid Inflatable Boat.

sastrugi: wavelike ridges of hard snow formed on a level surface by the wind.

serac: a pinnacle or ridge of ice on the surface of a glacier.

sit rep: situation report, a weekly briefing given by the base commander on base matters.

smoko: mid-morning and mid-afternoon tea break. A Navy term.

ACKNOWLEDGEMENTS

This book has had a long genesis. While working on it, I have been the fortunate recipient of several grants and awards. I wish to thank the following organizations, institutions, and ships for their generous assistance in the research and writing of *Ice Diaries*: the British Antarctic Survey; Arts Council England; the Shackleton Scholarship Fund; the Royal Literary Fund; the Canada Council for the Arts; the Natural Environment Research Council; the Environment Institute at University College London; the A.W. Mellon Foundation in South Africa; Galleri Svalbard in Spitsbergen, Norway; and the officers and crews of the RRS *James Clark Ross*, especially Jerry Burgan, and RRS *Ernest Shackleton*. These organizations also gave me valuable assistance and support in the writing of three other titles: *The Ice Diaries — Antarctic Work in Progress* (2006), *The Ice Lovers* (2009), and *Night Orders: Poems from Antarctica and the Arctic* (2011).

I would also like to thank the friends who kept me company in the Antarctic, the Arctic, in London, and in Norwich during the writing of this book: Layla Curtis, Gabriele Stowasser, Jerry Burgan, Silje Hørunges, Julia Bell, and Rachel Sieder. Henry Sutton has been a stalwart and supportive colleague at the University of East Anglia, as well as a perceptive reader of my work. Andrew McNaughton has provided a home, stability, and support in tropical Kenya, where much of this book was written and edited. As always, my special thanks goes to Diego Ferrari, photographer, collaborator, friend, fellow traveller.

This book brings together the expertise of many scientists and logistics people who spoke to me in the Antarctic, including many knowledgeable experts at the British Antarctic Survey, the Scott Polar Research Institute, Durham University, the University of Cambridge, Edinburgh University, the University of Southampton, and the University of East Anglia. Special thanks go to David Walton at the British Antarctic Survey and to Mark Maslin, professor of Climate Studies at University College London, for his collegial support and the clarity and concision of his published works on climate change.

I read many excellent books on the history of exploration and the polar regions generally in the research and writing of *Ice Diaries* and would like to acknowledge in particular the ones I quote from: *The Spiritual History of Ice: Romanticism, Science, and the Imagination* by Eric G. Wilson (New York: Palgrave Macmillan, 2003); *The Ice* by Stephen J. Pyne (London: Weidenfeld & Nicolson, 2003); *I May Be Some Time* by Francis Spufford (London: Faber and Faber, 1996); *Farthest North* by Fridtjof Nansen (London: Constable, 1897); and *South* by Ernest Shackleton (London: Penguin Classics, 2004). The ice terms which preface each chapter of this book are taken from the World Marine Organization "Sea Ice Nomenclature," which can be accessed at aari.nw.ru/gdsidb/glossary/p1.htm.

An extract from this book won the *Prism International* creative non-fiction competition in 2012; I thank *Prism* for its support over the years. It is one of North America's best literary journals, and I feel very grateful to have been published in its pages.

The images reproduced here are mine unless otherwise indicated. My thanks go to Jerry Burgan, Donald Campbell, and Rob Smith; to the Scott Polar Research Institute for permission to reproduce the image from the Terra Nova expedition "Captain Scott and group taken on return from the Southern Party" by Herbert Ponting, and to Hereford Cathedral for permission to reproduce the image of the *Hereford Mappa Mundi*.

I would like to thank my agent, Veronique Baxter at David Higham Associates in London, for her sound editorial advice, her tenacity, and commitment to my work. Susan Renouf at ECW Press has been the best editor possible — knowledgeable, insightful, and skilled. I am lucky to have found her. Thanks also go to the enthusiastic and efficient team at ECW: David Caron, Jen Knoch for her superb copy edit, Erin Creasey, and Crissy Calhoun. Natalie Olsen has created an elegant cover, which expresses so well the dual character of ice, that of serenity and threat. Finally, for the support, time, and space to write in South Africa, I wish to thank the late Stephen Watson, Steven and Denise Boers, the A.W. Mellon Foundation, Pieter le Roux, and Meg Vandermerwe for their generous friendship, and for giving me a lasting home in Cape Town.

Epilogue opener "Adelie penguin on an iceberg" (340–341) by ravas51 (flickr.com/photos/38007185@N00).

Cracked ice texture background (throughout) by Ian Mackenzie (flickr.com/photos/madmack).

Published by ECW Press
665 Gerrard Street East, Toronto, ON M4M 1Y2
416-694-3348 / info@ecwpress.com

To the best of her abilities, the author has related
experiences, places, people, and organizations
from her memories of them. In order to protect
the privacy of others, she has, in some instances,
changed the names of certain people and details of
events and places.

Library and Archives Canada
Cataloguing in Publication

McNeil, Jean, 1968-, author
Ice diaries : an Antarctic memoir / Jean McNeil.

Issued in print and electronic formats.
ISBN 978-1-77041-318-4 (bound)
ISBN 978-1-77090-875-8 (pdf)
ISBN 978-1-77090-876-5 (epub)

1. McNeil, Jean, 1968- —Travel—Antarctica.
2. Authors, Canadian (English)—Travel—
Antarctica. 3. Authors, Canadian (English)—
20th century—Biography. 4. Ice—Antarctica.
5. Ice—Social aspects. 6. Ice—Psychological
aspects. 7. Antarctica—Description and travel.
I. Title.

PS8575.N433Z85 2016   C813'.54   C2015-907305-7
C2015-907306-5

Editor for the press: Susan Renouf

Jacket design: Natalie Olsen | kisscut design   Author photo: Layla Curtis

Cover image: © schnee von gestern/photocase.com   Page design: Rachel Ironstone

The publication of *Ice Diaries* has been generously supported by the Canada Council for the Arts
which last year invested $153 million to bring the arts to Canadians throughout the country, and by the
Government of Canada through the Canada Book Fund. *Nous remercions le Conseil des arts du Canada
de son soutien. L'an dernier, le Conseil a investi 153 millions de dollars pour mettre de l'art dans la vie des
Canadiennes et des Canadiens de tout le pays. Ce livre est financé en partie par le gouvernement du Canada.*
We also acknowledge the Ontario Arts Council (OAC), an agency of the Government of Ontario, which
last year funded 1,709 individual artists and 1,078 organizations in 204 communities across Ontario, for
a total of $52.1 million, and the contribution of the Government of Ontario through the Ontario Media
Development Corporation.

Printing: Friesens 1 2 3 4 5   Printed and bound in Canada